# AGROFORESTRY
## A Sustainable Land Use System

## THE EDITORS

**Dr Raj Kumar,** Ph.D, Scientist (agroforestry) at ICAR- Central Soil Salinity Research Institute, Karnal-132001, Haryana, India. He has more than 8 years' research experience in the field of forestry and agroforestry. He is specialized in agroforestry, carbon sequestration, climate change and ecological engineering. He has published more than 30 research papers in reputed national and international Journals. Beside this, he has published three books, five book chapters, three training manuals and twenty abstracts in different national and international conferences.

**Dr Rakesh Banyal,** Ph.D in Forest Science with specialization in silviculture has more than 15 years' experience in research, teaching and extension in the field of forestry and/or agroforestry. His current research areas are development of climate resilient agroforestry systems, reclamation of salt affected lands and introduction of alternatives to saline waterlogged soils and addressing climate change issues through tree plantations. His past experience was on nursery techniques in forest trees species, short rotation forestry and plantation techniques on degraded sites. He has more than 70 research papers in journals of repute, popular articles in leading magazines, practical manuals, training compendium and book chapters in his credit.

**Manish Kumar,** Scientist (Agroforestry) at ICAR-Central Soil Salinity Research Institute, Karnal-132001, Haryana, India. He has more than 5 years' experience in the field of Agroforestry and related sciences. He is specialized in the agroforestry, tree-crop interaction, climate change adaptation and watershed management. He has published more than 8 research papers in reputed national and international Journals, five book chapters and fifteen abstracts in national and international conferences.

**Dr Sneha Dobhal,** Ph.D, assistant professor (forestry) at Veer Chandra Singh Garhwali Uttarakhand University of Horticulture & Forestry, College of forestry, Ranichauri. She has more than 2 year's research and teaching experience in the field of forestry. She is specialized in the fields of silviculture and tree improvement. She has published more than 10 research papers in reputed national and international journals.

# AGROFORESTRY
## A Sustainable Land Use System

Raj Kumar

Rakesh Banyal

Manish Kumar

Sneha Dobhal

2020

Associated Publishing Company®

*A Division of*

Astral International Pvt. Ltd.

New Delhi – 110 002

ISBN: 9789390371440 (Int. Edition)



*Published by* : **Associated Publishing Company®**
*A Division of*
**Astral International Pvt. Ltd.**
– ISO 9001:2015 Certified Company –
4736/23, Ansari Road, Darya Ganj
New Delhi-110 002
Ph. 011-43549197, 23278134
E-mail: info@astralint.com
Website: www.astralint.com

*Digitally Printed at* : Replika Press Pvt. Ltd.

# Preface

Agroforestry is a land use practice in which woody perennials (trees or shrubs) are integrated with crops and/or livestock on the same unit of land. In our country, area under agroforestry is estimated to be 17.45 M ha, while, 45% area has the potential to be brought under agroforestry. Moreover, in India, approximately 120 m ha land is subjected to various forms of degradation. Restoring these lands is extremely important to conserve natural resources, maintain landscape sustainability, mitigate climate change, improve socio-economic conditions and to ensure food security. Agroforestry has demonstrated its potential to restore degraded lands, reduce soil erosion, improve soil properties and to provide diversified income for the farmers.

Modern agriculture cover a range of risks related to production, prices and markets, financial, institutional and social aspects that often are directly or indirectly associated with weather impacts. In addition, increasing frequency, intensity and duration of severe weather events are posing major challenges to global food security and livelihoods of the rural people. The studies have shown that by optimizing sequential cropping patterns, the projected yield losses under climate change scenarios can be halved. In contrast, the buffers provided by trees and forests during periods of food insecurity may be adversely affected by climate change impacts and environmental degradation. Further, predictions suggest that climatic conditions will become more variable and extreme throughout the globe during 21$^{st}$ century. In response so-called 'climate-smart agriculture' solutions are intended to increase "resilience" to climatic impacts, improve livelihoods and food security, as well as addressing adaptation and mitigation objectives. One example of climate smart practice is agroforestry, whereby the deliberate addition of trees on farms is expected to sequester carbon while providing protection to adjacent crops, such as shading, wind-break and binding erosive soils. Agroforestry is climate resilient systems that have demonstrated great potential in adaptation and mitigation to climate change. It plays important role in climate moderation, reducing the impact of extreme weather events and enhancing the $CO_2$ sequestration.

Government of India has launched various programmes to double farmers' income by 2022. This involves reducing farm input and their cost, improving farm productivity, value addition and diversified farming etc. Agroforestry is one of the diversified farming systems that can improve farmer' income by growing high value tree such as teak and sandalwood in their farmland. In this context government has launched national agroforestry mission and bamboo mission to promote agroforestry among the farmers.

This book covers basic and applied aspect of agroforestry and is intended for the scientist, professors, experts and students those are working in the field of forestry, agroforestry, environmental science and ecology, climate science, policy planning and land use system designing..

**Raj Kumar**

**Rakesh Banyal**

**Manish Kumar**

**Sneha Dobhal**

# Contents

# List of Contributors

**Aman Dabral**

PhD Scholar, Genetics and tree improvement division, Forest Research Institute, Dehradun

**Abhishek Kumar**

Scientist, ICAR - Central Arid Zone Research Institute, Research Centre, Jaisalmer, Rajasthan

**Deepshikha Nirala**

PhD Scholar, Department of Silviculture and Agroforestry, College of Forestry, Dr. Y.S. Parmar University of Horticulture and Forestry, Nauni- Solan- 173 230, Himachal Pradesh

**Dhanush C. Reddy**

PhD Scholar, Department of Forestry and Environmental Sciences, GKVK Bengaluru

**Bharti**

Teaching Assistant (Statistics), Department of Basic Sciences, Dr. Y.S. Parmar University of Horticulture and Forestry, Nauni- Solan- 173 230, Himachal Pradesh

**Kanica Chauhan**

Assistant Professor, Department of Forest products and Utilization, College of Horticulture and Forestry Jhalarapatan, Jhalawar-326023 Rajasthan

**Nidhi Rawat**

Assistant Professor (Environmental Science), Uttaranchal (P.G.) College of Biomedical Sciences & Hospital, Dehradun

**Manish Kumar***

Scientist (Agroforestry), ICAR- Central Soil Salinity Research Institute, Karnal-132001, Haryana, India

**Pawan Kumar Poonia**

Assistant Professor, Department of Forest products and Utilization, College of Forestry (UAS Dharwad) Sirsi, Dist Uttara Kannada -581401 Karnataka

**Poonam**

Assistant Professor, Forest Biology and Tree Improvement, College of Forestry, Banda University of Agriculture and Technology Banda UP-210001

**Raj Kumar**

Scientist (Agroforestry), ICAR- Central Soil Salinity Research Institute, Karnal-132001, Haryana, India

**Sneha Dobhal**

Assistant Professor, VCSG Uttarakhand University of Horticulture and Forestry, College of Forestry, Ranichauri-249 199, Uttarakhand, India

**Sudipta Ramola**

Postdoc Fellow, Research Group for Advanced Materials & Sustainable Catalysis (AMSC), State key Laboratory Breeding Base of Green Chemistry- Synthesis Technology, College of Chemical Engineering , Zhejiang University of Technology, Hangzhou 310014,China

**Yourmila Kumari**

Senior Research Fellow, Department of Forest Products, Dr. Y.S. Parmar University of Horticulture and Forestry, Nauni- Solan- 173 230, Himachal Pradesh

**Manish Kumar**

Assistant Professor (Forestry), Doon Business School, Selaqui- 248001- Dehradun

**Mukesh Topwal**

Deputy Project Director ATMA, Department of Agriculture , Vikas Bhawan , Ladari, Uttarkashi, Uttarakhand,India-249193

**Pradip Kumar Sarkar**

Scientist, ICAR RCER, Research Centre, Ranchi, Jharkhand (India)

**Prshanth G. Naik.**

VCSG Uttarakhand University of Horticulture and Forestry, College of Forestry, Ranichauri-249 199, Uttarakhand, India

**Priyanka Chandra**

Scientist (Microbiology), ICAR- Central Soil Salinity Research Institute, Karnal-132001, Haryana, India

**S. Sarvade**

Assistant professor, College of Agriculture, Balaghat-481331, JNKVV, Jabalpur (MP), India

**Himanki Dabral**

Assistant Professor, Department of Agriculture, Shri Guru Ram Rai Institute of Technology and Sciences, Shri Guru Ram Rai University, Patel nagar, Dehradun, Uttarakhand (248001)

**Kamini Gautam**

Scientist, ICAR-Indian Grassland and Fodder Research Institute, Jhansi, UP - 284001, India.

**Shivaputra Bammanahalli**

College of Forestry, Dr. Y. S. Parmar University of Horticulture & Forestry, H.P, India

# Chapter - 1

# Introduction to Agroforestry

*Yourmila Kumari*

## Introduction

Social forestry pertains to those areas and forest which are manmade. Agroforestry is conspicuously and important part of Social forestry and is a multiple system of production i.e. tree and food crops, fodders, medicinal plant etc. Agroforestry is defined as sustainable land management systems which increase the overall yield of land, combined with the production of crops (including tree crops) and forest plants and animal simultaneously or sequentially on the same unit land and applies management practices that are compatible with the cultural practice adopted by the local population. Thus agroforestry is the co-existence of farm and forestry which is adopted on a scientific basis and consequently total yield of land is significantly increased. The Present status of land use pattern in India is as follows:

| | |
|---|---|
| Total land area | 329 m. ha. |
| Area under Agriculture | 143 m. ha. (47%) |
| Area under Forest | 75 m. ha. (22.7%) |
| Barren Land | 21 m. ha. |
| Under non Agriculture | 18 m. ha. |
| Illegally occupied | 24 m.ha. |
| Population | 1000 M. |
| Cattle production | 400 m |

- Half of the Forests in India are denuded to the various degrees due to increased human activities. It has been internationally acknowledged that 30 to 33% of the total geographical area must be under good forest cover for environment and ecosystem balance.
- India's fast growing population stands at a count of more than 1000 million. and cattle population about 400 million for which the huge amount of food, fodder, timber, fuel, medicines and employment is required.

- All the above situation calls for massive programme of Afforestation and planting with people's participation. This programme should attempt is restore ecological balance and meet the various needs of rural people. This is feasible only if tree growing become a people's programme which brought to be combined with agriculture.
- The land area of 75ha is under forest which is not adequate to maintain good environment. It was therefore, rightly resolved through a National Policy Resolution in the year 1952 to add 35 m ha to the forest cover and to bring 33% of our land under forests. The decision though wise, timely and far-sighted, was never implemented with the same spirit.

Against this background, Agroforestry should become an important land use system, conventionally which was duly recognized by planners while preparing the seventh plan document. At this stage a recommendation was also made that Agroforestry might be included as core subject in the curriculum by all the State Agricultural Universities.

## Historical Background of Agroforestry

Agroforestry has its root in many traditional farming practices, although it has only been recognized as a separate "technology" in the last two decades. For example, traditional shifting agriculture, still common in many parts of the world, makes use of the forest to rehabilitate the soil periods of crop production. In Indonesia, there are large areas under agroforestry developed by subsistence farmers where the fellow period is lengthened by enriching it with commercially important trees species. Moreover, the improved fellow techniques have built upon this same concept. Similarly, homestead orchard in temperate area which are often grazed underneath, have been a long tradition. The homegarden, seen in many tropical areas, is an ecological more complex extension of this practice with its multilayered combination of a variety of trees and crops, sometimes with animals. The taungya system was begun in Myanmar in 1806 as an aid establishing teak (*Tectona grandis*) plantations. Since then the concept of planting crop in the early stages of plantations have spread worldwide, and it has become recognized as a land management system.

In agriculture, scientific attention was focused on multiple cropping in the late 1960s and early 1970s. This, coupled with an awareness of some of the difficulties associated with modern, intensive, high input agriculture development, particularly in the tropics, lead to a search for other alternatives by the food and agriculture organization (FAO), the world bank, and others. Farmers and society in many countries have experienced problems such as massive soil erosion, loss of stream quality, and salinization, all of which rely on trees and or shrubs in order to overcome these problems. In Europe, a region for agroforestry development has been agricultural policy and the desire to set aside land from agriculture.

However, perhaps the most important impetus to the development of agroforestry was the report commissioned by the International development research centre (IDRC) of Ottawa, Canada. The team, led by John Bene, concluded in their 1977 report, that research priority should be given to the tropical forestry-agriculture interface as way to assist the poor in developing country. The report also recommended the establishment of an international organization to support this work. In 1978 the international council for research in agroforestry (ICRAF) was established within the consultative research group for international agriculture research (CGIAR). ICRAF, which in 1991 changed its name to the international centre for research in agroforestry and later on renamed as world agroforestry center, is based in Nairobi, Kenya, and has been playing a key role in the policy planning and development of agroforestry. It has moved from an initial emphasis on coming to terms with the myriad of traditional agroforestry practices, to providing many of the underlying science and social science concepts. In both the developing and industrialized world, agroforestry is now accepted as an important land use system, driven by the need to create sustainable and robust agroforestry ecosystems.

## Concept

Although its name suggests a new agricultural system, agroforestry is actually one of the oldest and most successful agricultural practices around the world. It is the result of generations of trial and error and depends heavily on the farmer's knowledge of the land, plants and the animals. But in an agriculture focused on higher production, agroforestry and diverse crop production have largely been replaced with monoculture.

Agroforestry is the combination of agricultural practices with forestry practices. It has become important for sustainable land-use in many parts of the world, although most clearly so in tropical parts of the developing countries. In developing countries it is seen as a means of poverty alleviation, particularly for rural peoples. The scientific application is relatively new, although many of the practices are ancient. Agroforestry now has a sound basis as an applied ecology, because of the very wide range of combinations of trees, understory plants, and perhaps animals, the actual implementation of practices is still dependent on careful observation and interpretation of the local situation and the needs of the farmer. Agroforestry is a collective name for land-use systems involving trees combined with crops and/or animals on the same unit of land. It combines:

- Production of multiple outputs with protection of the resource base;
- Places emphasis on the use of multiple indigenous trees and shrubs;
- Particularly suitable for low-input conditions and fragile environments;
- It involves the interplay of socio-cultural values more than in most other land-use systems; and
- It is structurally and functionally more complex than monoculture.

## Definitions

The term agroforestry is an "umbrella" term for those land-use practices and technologies where trees or other woody perennials are deliberately grown with crops, pastures or animals on farms.

Agroforestry is defined as the sustainable land-use system that maintains or increases total yields by combining food crops (annuals) with tree crops (perennials) and/or livestock on the same unit of land, either alternately or simultaneously, using management practices that suit the social and cultural characteristics of the local people and the economic and ecological conditions of the area.

Agroforestry is a collective name for a land-use system and technology whereby woody perennials are deliberately planted on the same land management unit as agricultural crops and/or animals in some form of spatial arrangement or temporal sequence. In an agroforestry system, there are both ecological and economical interactions between various components.

**According to the USDA**, the most basic definition is, "the intentional growing of trees and shrubs in combination with crops or forage. Agroforestry also includes tree and shrub plantings on the farm or ranch that improve habitat value or access by humans and wildlife, or that provide woody plant products in addition to agricultural crops or forage. Agroforestry is distinguished from traditional forestry by having the additional aspect of a closely associated agricultural or forage crop."

**The International Center for Research in Agroforestry** (ICRAF) defines agroforestry as a dynamic, ecologically based, natural resources management system that through the integration of trees on farms and in the agricultural landscape diversifies and sustains production for increased social, economic and environmental benefits for all land users at all levels. Thus, agroforestry focuses on the integration of farming and trees on a property, whether that property is a small home garden, a large estate or communally owned land. Agroforestry practices are also increasingly seen as a way of diversifying agro ecosystems by creating new niches for flora and fauna.

## Objectives

Objectives of agroforestry includes; revenue generation, conservation of natural resources, ecosystem services, ecosystem restoration, increased efficiency, economic diversification and/or moderation of financial risk. Indicators of success are determined according to the management objectives of the site.

## Main Features

The main feature of the multistoried production of different species of economic plants with suitable plant geometry. Agroforestry is **6-F** programme because it provides i.e. **food** for man, **fodder** for livestock, **fibre** for clothing, **fuel** for the village and urban homes, **furniture** from timber and **funds** for the poor forest villagers.

## Characteristics of Agroforestry Practices

- Land-use practices deliberately integrated into the whole farm.
- Contain complex interactions among components suited to particular environments and human needs.
- Competition and its management are critical.
- Agroforestry has two or more outputs.
- The "cycle" of an agroforestry practice is always more than one year.

- Even the simplest agroforestry practice is more complex, ecologically (in terms of structure and function) and economically than monocropping.
- This emphasis on the use of multipurpose trees and shrubs.
- It is structurally and functionally more complex system.
- Crops are arranged in such a way that it reduces the competition among plants for nutrient, air and light.

## Benefits of Agroforestry

### A) Environmental benefits

i). Reduction of pressure on natural forests.

ii). More efficient nutrient recycling by deep rooted trees.

iii). Better protection of ecological systems.

iv). Reduction of surface run-off, nutrient leaching and soil erosion through impeding effect of tree roots and stems on these processes.

v). Improvement of microclimate, such as lowering of soil surface temperature and reduction of evaporation of soil moisture through a combination of mulching and shading.

vi). Increment in soil nutrients through addition and decomposition of litterfall.

vii). Improvement of soil structure through the constant addition of organic matter from decomposed litter.

### B) Economic benefits

i). Increment in an outputs of food, fuel wood, fodder, fertilizer and timber.

ii). Reduction in incidence of total crop failure, which is common to single cropping or monoculture systems.

iii). Increase in levels of farm income due to improved and sustained productivity.

### C) Social benefits

i). Improvement in rural living standards from sustained employment and higher income.

ii). Improvement in nutrition and health due to increased quality and diversity of food outputs.

iii). Stabilization and improvement of communities through elimination of the need to shift sites of farm activities.

## Limitations of Agro-forestry

Agro-forestry does have certain limitations:

- Damage to food crops during harvesting of trees.
- Possible competition of trees with food crops for space, sunlight, moisture and nutrient which may reduce crop yield.

- Potential of trees is serving as hosts to insects and birds.
- Rapid regeneration of profile trees may displace food crops and take over entire fields.
- The interference of trees decreases the crop yield which is lower than monocropping.
- Competition for moisture between tree and arable crop is minimum when the trees have deep root system with less lateral roots. Depletion of soil moisture is higher under trees having lateral root system (Eucalyptus) and higher water requirement (Bamboo).
- Competition for nutrients occurs when the trees having finer roots in the top soil.
- The felling of trees may cause damage to arable crops.
- Some trees serves as a hosts to pests that harms food crops.
- Agroforestry system requires more labour.
- Longer gestation period for trees delays the returns to farmers
- Farmers' resistance to replace the food crops with trees is common.
- Birds' attraction by trees causes increased damage to arable crops.
- Certain tree crops release chemicals (allelopathy) that adversely inhibit arable crops (Zuglans, Eucalyptus).

Through skilled management practices any or all these aspects can be controlled. For example, once it is known that trees compete with food crops and may reduce food yields, it is easy to adopt some of the following strategies.

- Select legume trees that have small or light crowns so that sunlight will reach the food crops.
- Select trees that are deep-rooted so that they will also absorb moisture and nutrients from the deeper subsoil.
- Space the trees further apart to reduce their competitive effect on the food crop.

## Scope of Agro-forestry In India

There is tremendous scope for Agroforestry because India has achieved self-sufficiency in food production. Now its attention is becoming more focused on the ecological problems and shortage of fuel, fodder and other outputs as well as unemployment. Agroforestry has vast scope in meeting this requirement through multipurpose tree species as:

i). large area is available in the form of farm boundaries, bunds, waste lands where this system can be adopted

ii). This system permits the growing suitable tree species in the field where most annual crops are growing well iii). By growing trees and crops on agricultural or forest land, resources are utilized efficiently

iv). System has potential to generate employment.

v). Provides raw material for the cottage industries

vi). Trees help in maintaining ecological balance

vii). Trees help in soil and water conservation, and soil improvement.

viii). Trees help in meeting various needs of growing population.

**Components of agroforestry:** In agroforestry systems three basic sets of components are managed by the land user, i.e tree or woody perennial, crops/herb (agricultural crops including pasture species), and the animal. A land-use system to be designated as an agroforestry system, it must always have a woody perennial. In most of the agroforestry systems, the annual or herbaceous species is also involved, except in case of apiculture and aquaculture with trees, and plantation-crop mixtures of two woody perennials such as coffee and rubber trees, or coffee, cacao, and tea under shade trees. Animals are only present in some agroforestry systems such as silvipastoral systems. Overall there are only three basic components of agroforestry:

1. Agricultural crops/herbaceous plants
2. Woody perennials (tree crops/forest plants)
3. Animals.

The above basic components when combine together form three basic agroforestry systems.

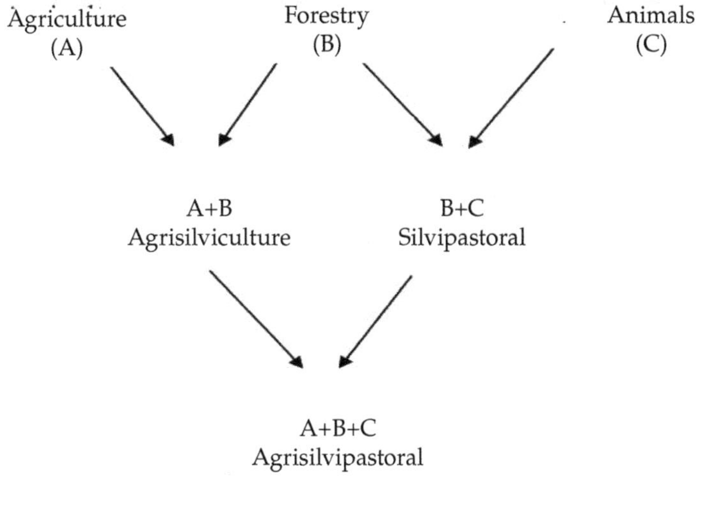


Interaction of various components in agroforestry systems

In general there are 4 or 5 basic sets of components which are managed by human beings in all agroforestry systems. Structurally, this system can be grouped as:

1. Agri-silviculture system
2. Agri-horticulture system
3. Silvipastoral system
4. Agri-silvipastoral system

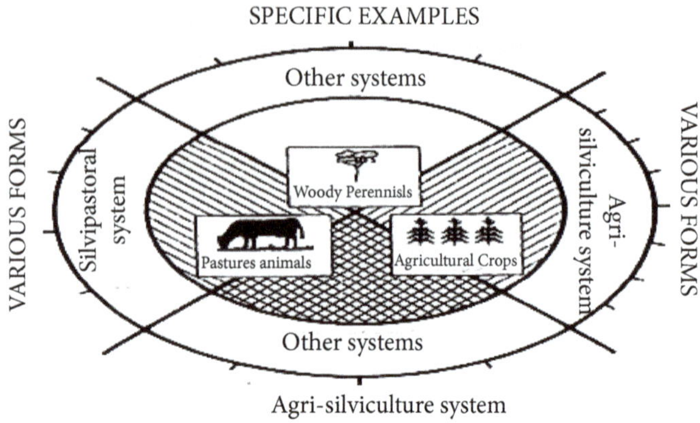


**Figure 1. Components of agroforestry (Nair 1985).**

Agroforestry systems in detail can be classified based on structure, function, agro ecological adaptation and technological inputs.

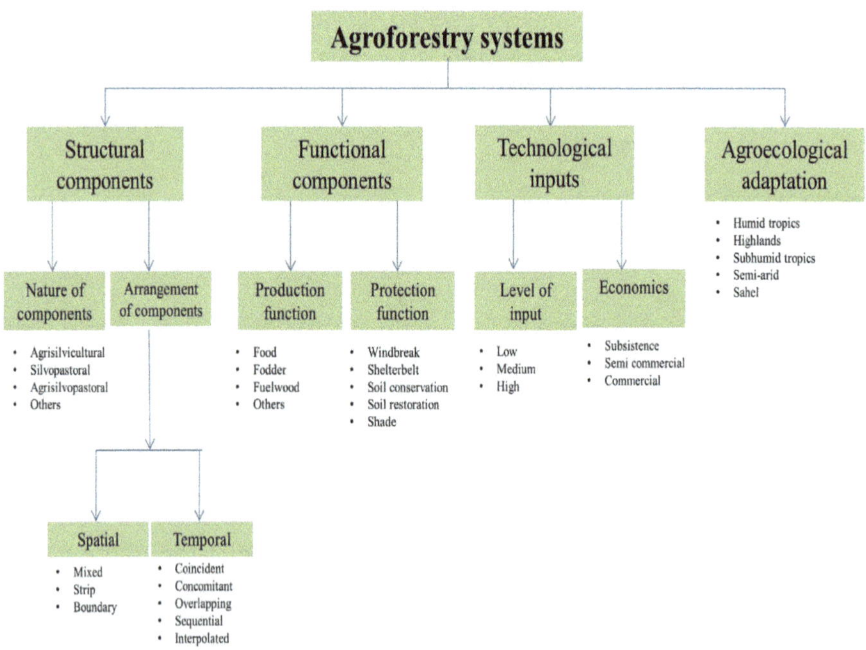

**Figure 2. Detailed classification of agroforestry systems**

On the basis of nature of components the following are common agroforestry systems found in different agroecological regions of India:

1. Agrisilviculture (trees + crops)
2. Boundary plantation (tree on boundary + crops)
3. Block plantation (block of tree + block of crops)
4. Energy plantation (trees + crops during initial years)
5. Alley cropping (hedges + crops)
6. Agri-horticulture (fruit trees + crops)
7. Agri-silvihorticulture (trees + fruit trees + crops)
8. Agri-silvipasture (trees + crops + pasture/animals)
9. Silvi-olericulture (tree + vegetables)
10. Horti-pasture (fruit trees + pasture or animals)
11. Horti-olericulture (fruit tree + vegetables)
12. Silvi-pasture (trees + pasture/animals)
13. Forage forestry (forage trees + pasture)
14. Shelter-belts (trees + crops)
15. Wind-breaks (trees + crops)
16. Live fence (shrubs and under- trees on boundary)
17. Silvi-horti-sericulture (trees/fruit trees + sericulture)
18. Horti-apiculture (fruit trees + honeybee)
19. Aqua-forestry (trees + fishes)
20. Homestead (multiple combinations of trees, fruit trees, vegetable etc).

## Spatial Arrangement of Trees in Agroforestry

Spatial arrangement of tree in also one of the important component of agroforestry. In order to decide what arrangement would work best as well as the species, we need to take into account farm boundaries and natural features such as road, rocks, springs and streams. On steep slopes make terraces or contour strips (of trees, grasses and/or other plants) to control and avoid erosion. The interactions of trees with agricultural crops and livestock are important in planning agroforestry. Interactions between tree can be positive (e.g. increased fertility, protection) or negative (e.g. suppressed growth, competition for resources). The productivity of an agroforestry system depends on these interactions. It is beneficial to adopt arrangements with high tree -crop interface if the resultant interactions are positive; adopt arrangements with low tree – crop interface if interactions are negative but both components are still required e.g. for environmental services. Benefits will only be obtained through a combination of the right tree species with the right crops in the right spatial arrangements and with the right management practices. The various types of arrangement of trees in agroforestry has been shown in figure 3.

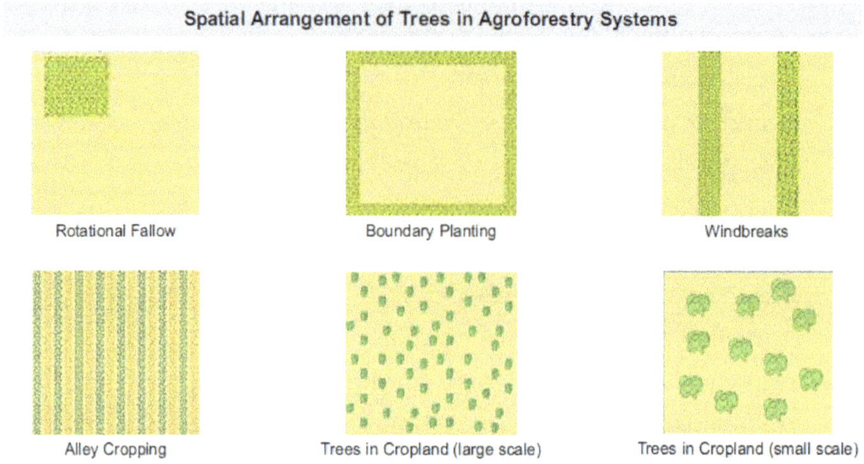


**Figure 3. Spatial arrangement of trees in agroforestry systems**

## Agroforestry Systems Distribution in India

In India, various forms of agroforestry systems are prevalent in different agro-climatic regions. These agroforestry systems occupy sizeable area in the country with all the systems scattered throughout the country. Geospatial technologies used for mapping and delineating agroforestry area of the country. Since agroforestry consist of both tree and crop components, so there are some problems in mapping agroforestry. According Rizvi et al., (2014), area under agroforestry is estimated to be 17.45 M ha in India. However, another estimate suggesting agroforestry area in India estimated to be 14.46 M ha without fallow & 17.45 M ha with fallow land as against 25.32 M ha (Dhyani et al. 2013) (figure 4).

## Potential Area for Agroforestry in India

Ahmed et al., (2018) analyzed the land potentiality in India for agroforestry suitability reveals 32.8% as highly suitable (S1), 40.4% moderately suitable (S2), 11.7% marginally suitable (S3), and 9.1% not suitable (NS). About 52% of land of India is under the cropping systems. In addition, it revealed that the 46% of these cropland areas fall into high agroforestry suitable category "S1," and provide huge opportunity for harnessing agroforestry practices. Furthermore, agroforestry suitability mapping in broad ecosystem and in different agroecological regions will assist various projects in India at the regional level (figure 5).

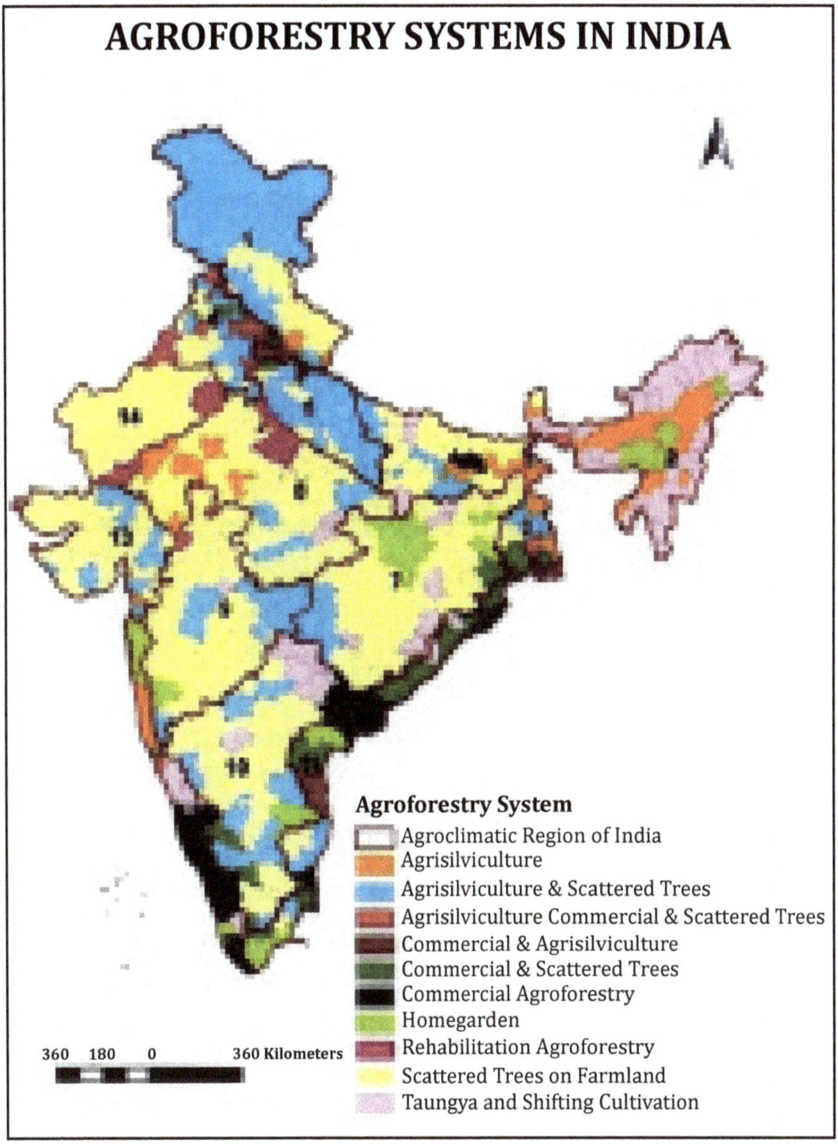


**Figure 4. Thematic map showing various agroforestry systems in India (Source: Rizvi et al., 2014)**

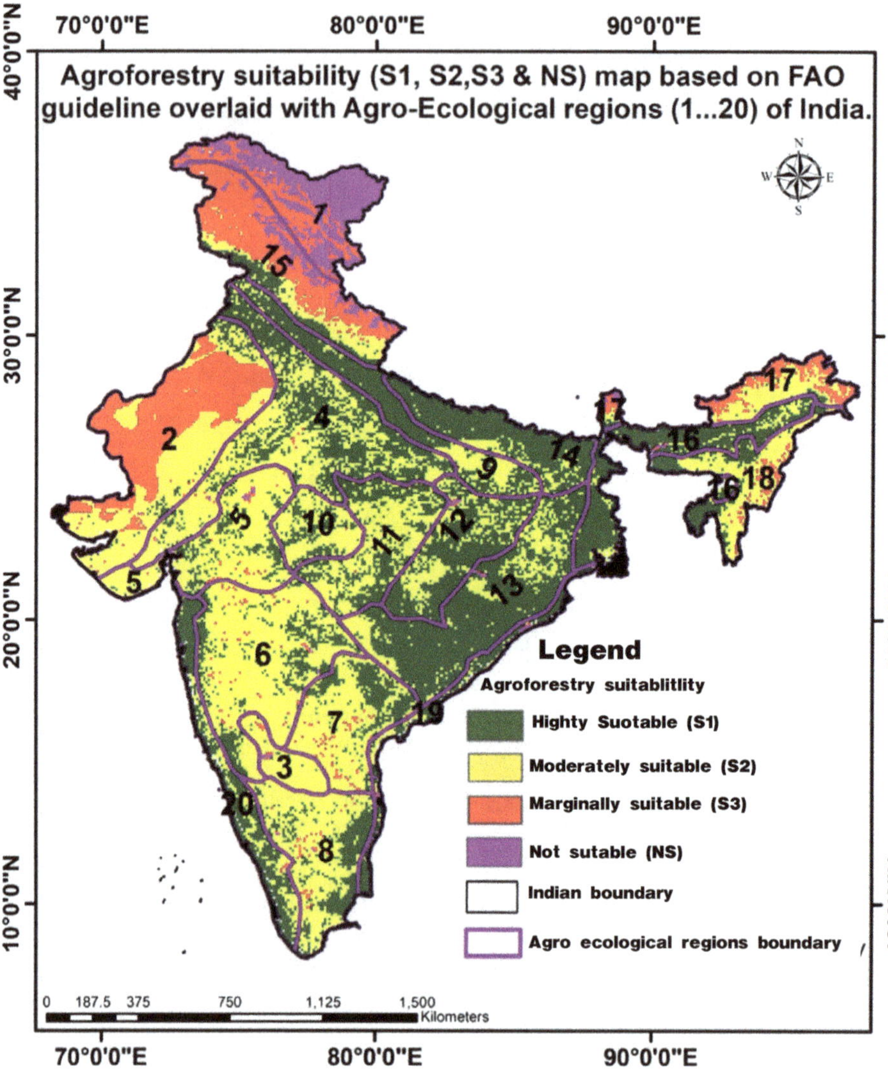


**Figure 5. Agroforestry suitability map of India (Source, Ahmed et al., 2018)**

## Summary

Agroforestry is a dynamic, ecologically based management system for growing woody perennials, either trees or shrubs, with crops and/or pastures and/or animals. There is a wide range of systems and practices, with their selection and use dependent on ecological, biophysical and social factors. The main focus of agroforestry is to meet the needs of rural people in developing countries in a sustainable manner. Its growth and implementation has been closely linked with social and community forestry programs in developing countries. However, it is also accepted as a viable land-use practice in industrialized countries. While agroforestry, as a subject, is relatively new, its roots and many of its practices are very old. Research and development initially focused on the classification, description and use of existing practices. Later the research focus shifted to understanding the complex ecological factors between trees, crops and animals. A recent emphasis is on viewing agroforestry from an ecological perspective, where complexity, biodiversity and succession are important attributes to make the systems sustainable and resilient. There are increasing efforts at domestication of agroforestry trees and in developing markets for their products, because profitability is a seen as key to sustainability and poverty alleviation. Understanding biophysical interactions between trees, crops, animals and the soil, along with the wider ecological view of their dynamic nature, is the basis for better implementation and the development of new agroforestry practices. The presence of trees and ecological complexity is important for ensuring biological sustainability. Agroforestry has often been promoted as being a sustainable solution to poverty alleviation and soil degradation problems and for use in soil reclamation projects. As practitioners, science and social science gain better insight into and experience with agroforestry, it will be applied more widely, carefully and with improved integration at a farm, catchment or landscape level.

# Chapter - 2

# Agroforestry: Past, Present and Future

*Poonam*

## Introduction

It has been the practice to cultivate tree species and agricultural crops in intimate combination since the dawn of civilization. It was the general custom in Europe until the Middle Ages to clear-fell forests, burn the slash, cultivate food crops for varying periods on the cleared areas and plant or sow tree species before, along with or after the sowing of the agricultural crop (Anon,1938 and Bene, 1977). In tropical America, farmers have long imitated the structure and species diversity of tropical forests by planting a variety of crops like coconut or papaya with a lower layer of banana or citrus, a shrub layer of coffee or cacao, tall and low annuals such as maize and finally a spreading ground cover of plants such as squash with different growth habits (Wilken,1977 and Blanford, 1958). In Asia, the Hanunoo practice of Philippines is a complex and somewhat sophisticated type of shifting cultivation. Clearing of forests for agricultural use in which they deliberately left certain selected trees in the end of the rice-growing season to provide a partial canopy of new foliage for prevention of excessive exposure to the sun at the time of grain filling and maturity (Conklin, 1953). In Zambia, in addition to the main crop numerous subsidiary crops were grown with tree species (Anon., 1938). In 1806, in Burma, established a plantation of teak through the use of what he called the "taungya" method and presented it to Sir Dietrich Brandis (Blanford, 1958). From these beginnings, the practice became more and more widespread. It was introduced into South Africa as early as 1887 (Hailey, 1957) and was taken from Burma to the in India in 1890. In 1920 it was adopted in Travancore (now Kerala), in 1923 in the United Province (now Uttar Pradesh) and in 1925 in the Central Provinces (now Madhya Pradesh) (Raghavan, 1960). However, little and no thought appears to address the drawbacks of the systems from farmers perspectives and his farm produce in between 1956 to 1970. The origin of agroforestry practices, i.e. growing trees with food crops and grasses, is believed to have been during Vedic era (Ancient period, 1000 BC), the agroforestry as a science is introduced only recently. The systematic research in agroforestry geared up after the establishment of the International Council for Research in Agroforestry (lCRAF) in 1977, which was renamed in 1991 as the International Centre for Research in Agroforestry. During 2001-02, ICRAF adopted a new brand name "World Agroforestry Centre", to more fully reflect their (ICRAF's) global reach and also their more balanced research and development agenda; however their legal name "International Centre for Research in Agroforestry" will remain unchanged. In India, the research on agroforestry was initiated in the late 1960 and 1970 by the Indian Grassland and Fodder Research Institute, Jhansi, Indian Institute of Soil and Water Conservation Research and Training, Dehradun, Central Arid Research Institute, Jodhpur and ICAR-Research Complex for North Eastern Hill Region. It is also documented in literature that even prior to this the forest departments had taken numerous studies on tangya cultivation. The National Commission on Agriculture emphasized agroforestry education in the 7th Five Yaer Plan period wherein all state agriculture universities have introduced it into the agriculture syllabus in accordance with the recommendation of the task force constituted during the first agroforestry seminar convened at Imphal, India in May, 1979. The Indian Society of Tree Scientists (ISTS) with its H/Q at UHF, Solan organized national seminar on 'Agroforestry a New Challenges' in 1985 and international seminar on 'Agroforestry for rural needs' in 1987. ICAR had launched an All India Co-ordinated Research project on Agroforestry located at CAFRI, Jhansi in 1983 with 22 centres and now spread to 35 coordinating centres network in 2018 which are operational in different agro-ecological region of the country. The mandate of the project is to screen and henetic upgrading of different plant species for their compatibility, to study the trees on legumes/trees/suitable orchard species while growing mixed with suitable species of grasses and fodder commercial plants and performance of the local predominant grasses/local crops and refining the technology for proper propagation and management of the species. During 7th Five Year Plan a post of Asstt. Director General (Agroforestry) was created at the ICAR headquarter in New Delhi to coordinate the research on agroforestry in India. The UNDP centre of advance studies on Agroforestry for post-graduate Education was established at the Dr YS Parmar University of Horticulture and Forestry, Nauni, Solan Himachal Pradesh was the another landmark in the history of agroforestry development and further a separate degree program has been started from August, 1992. Similar, program was incepted in other universities in states like Gujarat, Haryana, Uttar Pradesh and Jammu & Kashmir.

In developing countries like India, the demand for food and other things has increased many folds in the recent past. But, the arable land area in the country is finite and cannot be expanded further. India has about 329 mha land, of which 44 per cent of area is degraded (Mythili and Goedecke, 2016). On the other hand, the existing tree cover of Indian forests is not at par with the guidelines of National Forest Policy (NFP), 1988 wherein it is stated that '*minimum of one-third (33%) of the total land area must be under forests and tree cover to achieve the national goal for ecological security*'.

The current forest cover is 21.54 percent in the country consisting of very dense forest (2.99%), moderately dense forest (9.38%) and open forests (9.18%), respectively. The major chunk of Indian forests falls under moderately and open forest category of serious concern (DNFP, 2018 and FSI, 2017). The National Agriculture Policy (2000) emphasized the role of agroforestry for efficient nutrient cycling, nitrogen fixation, organic matter addition and for improving drainage and underlining the need for diversification by promoting integrated and holistic development of rainfed areas on watershed basis through involvement of community to augment biomass production through agroforestry and farm forestry. The Task Force on Greening India for Livelihood Security and Sustainable Development of Planning Commission (2001) has also recommended that for sustainable agriculture, agroforestry may be introduced over an area of 14 million ha out of 46 m ha irrigated areas, that are degrading due to soil erosion, water-logging and salinization. For integrated and holistic development of rainfed areas, agroforestry is to be practiced over an area of 14 million ha out of 96 m ha. This all will, besides ensuring ecological and economic development provides livelihood support to about 350 million people. The practice of agroforestry can help in achieving these targets. Therefore in the quest of optimizing productivity, the multi tier system came into existence. Gap of demand and supply of forest produce in India is widening and forests are unable to fulfill the demand. Agroforestry can play an important role in filling this gap and conservation of natural resources.

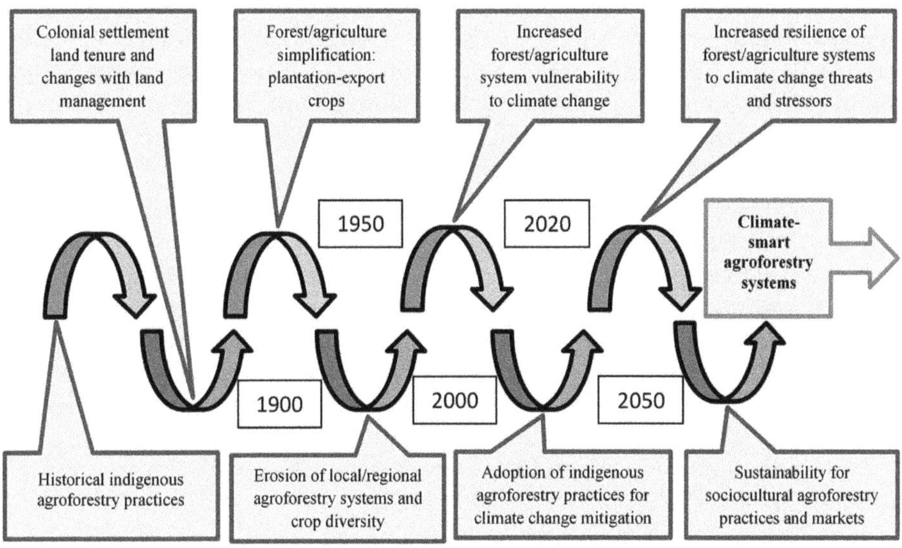


**Figure 6. Timeline depicting influences on indigenous agroforestry systems and their adaptive responses to change (Source: MacFarland, 2017)**

## Agroforestry History in India

In about 700 BC, man changed from a system of hunting and food gathering to food production. Shifting cultivation in India is prehistoric and partly a response to agroecological conditions in the region. Horticulture as co-existent with agriculture is found to have been prevalent in India from early historic period (500 BC to 1st century AN) when a certain amount of share in garden crops started to have been enjoyed by the king for providing irrigation. Luukkanen, *et.al.,* 2013). The cultivation of date-palm, banana pomegranate, coconut, jujube, aonla, bael, lemon and many varieties of other fruits and requirement of livestock in agriculture and mixed economy of agriculture and cattle-breeding may be traced in proto-history chalcolithic periods of civilization. (Sharma, 2007, Chavan *et al.,* 2014). But in India, the plant husbandry (intentional sowing or planting for production of desirable plants or plant domestication) happened to start under progressively arid climatic zone from about pre-Neolithic period (World Bank, 1984). The role of many common trees such as khejri or sami (*Prosopis cineraria*), aswattha (*Ficus religiosa*), palas (*Butea monosperma*) and varana (*Crataeva roxburghii*) in Indian folk-life has been mentioned in ancient literature of Rig Veda, Atharva Veda and other ancient scriptures (Dhiman, 2013). Traditional agroforestry systems manifest rural people's knowledge and methods to benefit from complimentary uses of annuals and woody perennials on the sustained basis (Zomer, 2009). It also indicates that farmers have a closer association with trees than any other social group and promoters of forests. In central America it has been a traditional practice since long by farmers to plant about two dozen species on a small piece of land. In Europe, until middle ages, it was the general custom to clear fell degraded forests, burn slash, cultivate food crops for varying periods on cleared area and plant trees before or along with, or after sowing agricultural crops. This farming system was widely practised in Finland up to the end of the last century and in a few areas in Germany as late as the 1920s. In certain eastern countries, people clear forest for agricultural use and deliberately spared certain trees which by the end of the rice-growing season provided partial canopy of new foliage to prevent excessive exposure of soil to sun.

In India, farmers and land owners integrate a variety of woody perennials in their crop and livestock production depending upon the agroclimates and local needs. Most of these practices are, however very location specific and information on these are mostly scanty. Therefore, their benefits have remained under exploited to other potential sites. It has now been well-recognized that agroforestry can address some of the major land use problems of rainfed and irrigated farming systems in India, and that a great deal can be accomplished by improving indigenous systems. With the current interests in agroforestry worldwide, attempts are being made in India to introduce agroforestry techniques using indigenous and exotic multipurpose and nitrogen-fixing woody perennials.

## Traditional Agroforestry Systems in India

Agroforestry is widespread in all ecological and geographical regions of India. The systems vary enormously in their structural complexity and species diversity, their productive and protective attributes and their socio-economic dimensions (Planning Commission, 2001). They range from apparently simple forms of shifting cultivation to complex home-gardens: from systems involving sparse stands of trees on farm lands (e.g. *Prosopis cineraria* khejri tree - in arid regions of western India) to high-density complex multistoried homesteads of humid lowlands: from systems in which trees play a predominantly 'service' role (e.g. shelter belts) to those in which they provide main saleble products (e.g. intercropping with plantation crops) (Dhyani, 2013).

## 1. Shifting cultivation (slash and burn system)

It refers to farming system in north-eastern high rainfall areas in which land under natural vegetation (usually forests) is cleared by slash and burn method and cropped with common arable crops for a few years and then left unattended when natural vegetation regenerates. Traditionally, the fallow period is of 10 to 20 years but in recent times it is reduced to 2 to 5 years (Planning Commission, 2001). Due to the increasing trends of population pressure, the fallow period is drastically reduced and system has degenerated causing serious soil erosion depleting soil fertility resulting to low productivity. In north-eastern India many annual and perennial crops with diverse growth habits are being grown. At times annual crops such as potato, rice, maize and ginger are grown in monoculture or mixed culture along with *Pinus kesiya*. Another important attribute of the system is secondary succession of vegetation during fallow period (Raghavan, 1960). The tree species which may be considered suitable for afforestation of fallow areas or to intercrop with arable crops must be fast growing preferably nitrogen-fixing and must efficiently recycle available nutrients within the system shortening time required to restore fertility. These may include species of *Acacia, Albizia, Alnus, Casuarina, Erythrina, Faidherbia, Gliricidia, Inga, Leucaena, Parkinsonia, Pilhecellobiuin, Prosopis, Robinia* and *Sesbania.*

## 2. Taungya system

The Taungya system is like an organized and systematically managed shifting cultivation. The word is reported to have originated in Myanmar (Burma) in 1850s (Blanford, 1958), made oup two words i.e. taung means hill and ya means cultivation (hill cultivation). It involves cultivation of crops in forests or forest trees in crop fields and was introduced to Chittagong and Bengal areas in colonial India in 1890. Later it had spread throughout Asia, Africa and Latin. The taungya system was introduced in India by Brandis in 1890 and the first taungya plantation was raised in 1896 in North Bengal. Further it was introduced to Coorg in Karnataka. Regular plantation however started in North Bengal in 1911 for raising Sal plantations and in 1912, extended for raising Teak. In 1923 it was adopted in UP for raising Sal plantations. In southern India, the system is called 'kumari'. It is practiced in areas with an assured annual rainfall of over 1200-1500 mm.

## 3. Home-gardens/homesteads Cultivation

A homestead is an operational farm unit in which a number of crops including tree crops are grown with livestock, poultry and/or fish production, mainly for the purpose of meeting the routine basic needs of the farmer (Roche,1976). It is an old-age practice in coastal states particularly in Kerala, Tripura, Assam, North-eastern states and parts of West Bengal and Andaman and Nicobar Islands (Tewari, 2007, Zomer, 2009). Athough the home-gardens appear to be a mixture of trees, shrubs and herbs, a certain general pattern seems to exist. There is wide variation in the intensity of trees species and crops based on the size of holding, needs of the people residing in homesteads and micro-climate (Nair,1979). Coconut, arecanut, guava, mango, citrus, tamarind, jack-fruit, papaya, banana, moringa, sesbania, custardapple and many multipurpose trees are the major trees found grown in home-gardens.

### 4. Plantation based Agroforestry systems

Modern commercial plantation crops like rubber, coffee, poplar, eucalypts and oilpalm represent a well-managed and profitable stable land-use activity in tropics. Some of the plantation crops like coconut, palms have been cultivated since very early times but their economic yield remained low. However, the research attention and commercial yields of these crops have increased substantially (Ram, 2005). Most of the coconut production in India, and other countries comes from small holdings, in which coconut-palm is integrated with a large number of annual and perennial crops like clove, cardamom, coffee, cacao, cassava, yams, fodder grasses and legumes (Tewari, 2007). Important cereals grown with coconut include rice, finger millet and maize, and pulses such as pigeon pea, green gram, black gram, cowpea, soybean, oilseed crop such as groundnut; root crops such as sweet-potato, yams, elephant foot-yam and taro. Spices and condiments such as ginger, turmeric, cinnamon, clove, chillies and black pepper. Fruits like pineapple, mango, banana and papaya and cash crops such as cotton and sugarcane, plantation crops such as areca nut, cacao and coffee. Many trees such as species of Erythrina, Ceiba, and Cordia find a place in the system (FSI, India 2013).

### 5. Scattered trees on farm lands

The practice of growing agricultural crops under scattered trees on farm lands is old and does not seem to have changed for centuries. Though world-wide list of such trees is long, some of them have received more attention than others e.g. *Prosopis cinaria* in western India and Poplars in north India (Chavan, 2014, Tewari, 2007). Some of the practices are very extensive and highly developed in India. For example growing *Prosopis cilleraria* and *Zizuphus* in arid areas; *Acacia nilotica* in Indo- Gangetic plains; *Grevia optiva* and other tree species in hills of Uttarakhand and HP; *Eucalyptus globulus* in southern hills of Tamil Nadu and *Brossus flabellifer* in peninsular coastal regions (Zomer, 2009).

### 6. Trees on farm-boundaries

Trees grown in agricultural fields are also often and usually grown on farm boundaries. In northern parts of India, particularly in Haryana and Punjab Eucalypts and Poplar are commonly grown along the field boundaries or bunds of paddy fields; other trees which are found grown as boundary plantations or live hedge include-. *Dalbergia sissoo* and *Prosopis juliflora* (Dhyani, 2013). Farmers' of Sikkim grow bamboo (Dendrocalamus, Bambusa) all along the irrigation channels. In coastal areas of Andhra Pradesh, *Borassus* is the most frequent palm. In Andamans, farmers grow *Gliricidia sepium*, *Jatropha* spp, *Ficus, Ceiba pentandra, Vitex trifolia and Erythina variegata* as livehedges (Chavan, 2014). At many places succulents like Agave and many cactoids are grown as common live-fence.

### 7. Wood lots

In many parts, farmers grow trees in separate blocks as wood-lots along with agricultural fields. Now the practice is expanding fast due to shortage of fuel-wood and demand of poles or pulp-wood in industry. For example, bamboo poles are in great demand for orange orchards in Nagpur area and Eucalvptus and Poplars for match industries (ICFRE, 2012). These days wood-lots are being raised mostly on large farms due to increase of labour costs and labour management, lack of irrigation and risk of crop investments. Wood-lots of Casuarina, bamboo, Poplar, *Eucalyptus, Leucaena leucocephala,* red sandal (*Pterocarpus santalinus*) and *Dalbergia sissoo* have become popular in many parts of the country (Luukkanen *et.al.*, 2013).

### 8. Systems for soil conservation or amelioration

About 150 million ha of land in India is subject to serious wind and water erosion, of which 69 million ha are critically affected. About 4 million ha is suffering from degradation due to ravines and gullies. In India, 6.75 m ha area is salt-affected out of which 2.92 and 3.83 m ha are saline and sodic, respectively. The menace, increasing with each passing year to the irrigated areas, has become a major concern in canal command areas. Agroforestry plays pivotal role in biological amelioration and check in further expansion of such landmasses. It is always useful to have agroforestry models/systems for saline landmasses by involving multipurpose nitrogen fixing tree species, fruit trees, halophytes and arable crops of economic importance. Coastal sandy areas and steeply sloping lands and more than 9 million ha is salt affected. The deep and narrow gullies are best controlled by putting them to permanent vegetation after closure to grazing. Afforestation with suitable tree species like *Acacia nilotica, Azadirachta indica, Butea moonosperma. Prosopis juliflora, Dalbergia sissoo, Tectona grandis, Bambulsa spp.* and *Dendrocalamus* and other adaptable species such as grasses like *Dichanthium annulatum, Bothriochloa pertusa, Cynodon dactylon* and *Sehima nervosum* will help in stabilizing ravines and gullies and checking their further spread. In recent times, due to increase in population pressure these wood-lots have shrunk at a fast rate.

### 9. Shelter belt

Arid regions witness very high wind velocity throughout the year. Farmers build shelter-belt (kana bundi) by either small dead wood or local vegetation to check wind velocity within safer limits. *Crotolaria burhia, Leptadenia pyrotechnica* and *Aerva pseudotomentosa* bushes are planted in 20-25 m a part rows across the wind direction. Between the lines of these shrubs grasses such as *Cenchrus ciliaris, C. setigerus, Lasiurus sinicus* etc are planted on leeward side of each break. This permanent vegetation helps accumulate sand near them which is again spread in the field. This also helps increase crop yields along the lines.

### 10. Trees on rangelands

As pointed out above, *Salvadora oleoides, Capparis decidua. Acacia nilotica, A. leucophloea. P. cineraria,* and now *Prosopis juliflora* also are most frequent trees on common community grazing lands. In tropics coconut is most common tree on pasture lands. Cattle raising usually involves grazing on these pastures. In some cases special fodder plants including legumes are also cultivated.

## Agroforestry in Recent Past

In the last 3-5 decades spectacular improvements have been made in tree plantation. The most important agroforestry practice is known from the Kangeyam tract of Tamil Nadu, where *Acacia leucophloea* + *Cenchrus setigerus* in silvi-pasture system was perfected. (Nair, 1979). Similarly, in ravines of Yamuna and Chambal, trees, shrubs and bamboos with grasses were planted for rearing milk-producing Jamunapari breed of goats and sheep. Scattered trees with khejri or mehndi in association with bajra, jowar and chillies were grown in the semi-arid area of Tamil Nadu (Benjamin, *et.al.*, 2000). Tree plantation continued as demarcation and control against wind erosion throughout the country. Plantation of khejri trees for various uses was common practice in Rajasthan. In coastal areas, *Casuarina equisetifolia* and other trees were grown in association with crops on farm lands for cash and to generate small timber (Mayrand, 2004).

## Scope of Agroforestry

Agroforestry applies to private agricultural and forest lands and communities. These are highly disturbed, human-dominated land-use systems. Targets include highly-erodible, flood-prone, economically marginal and environmentally sensitive lands. (King, *et.al.,* 1978). The typical situation is agricultural, where trees are added to create desired benefits. Our goal is to restore essential processes needed for ecosystem health and sustainability, rather than to restore natural ecosystems (Wilken, 1977). Agroforestry provides strong incentives for adoption of conservation practices and alternative land uses and supports a collaborative watershed analysis approach to management of landscapes containing mixed ownerships, vegetation types and land uses by:

1. Controlling poverty through increasing income by higher production of agroforestry products for home consumption and market.
2. Food security by restoring farm soil fertility for food crops and production of fruits, vegetables nuts and edible oils.
3. Empowerment to women farmers and other less-advantaged rural residents whose rights to land are insecure through better negotiations.
4. Reducing deforestation and pressure on forest by providing fuelwood grown on farms.
5. Increasing buffering capacity of farmers against the effects of global climate change on-farm tree crops and tree cover.
6. Improving soil health of the farm through ameliorated micro-climate effects.
7. Augmenting accessibility to medicinal trees for cure of common and complex diseases.

## National Agroforestry Policy, 2014

Agroforestry with significant contribution and potential has not gained the desired importance as a resource developmental tool due to its dependency on multi-institutions and multi-disciplinary approach. At the highest level, while the Ministry of Agriculture, Government of India has the mandate for agroforestry research, agriculture is a state subject and forestry is in the concurrent list and there are no extension services dedicated for its promotion. Besides, restrictions imposed by the State Government (as per Section-41 of India Forest Act) on felling and transportation of trees grown even on farmland, legal requirements for obtaining permits for felling and transportation, lack of institutional finances, absence of adequate infrastructure for marketing wood produce and lack of market information are some of the few hurdles which discourage the farmers from undertaking tree planting on farmland. There is a serious lack of institutional mechanisms to promote agroforestry at all levels. To overcome all these hurdles, India has announced the landmark National Agroforestry Policy 2014 that will mainstream the growing of trees on farms to meet a wide range of developmental and environmental goals. For planning and implementation, the Policy has recommended to develop a sound database and information system on agroforestry. However, at present unlike forest cover and area under crops, exact estimates for area under agroforestry are not available. The major problem in estimating the same lies in the different modes in which it occurs. The World Congress on Agroforestry held in New Delhi in February 2014 to have a forward outlook to any constraints that might restrict the adoption of agroforestry practices, wherein the President of India launched the much needed National Agroforestry Policy (NAP), 2014 first of its kind of path breaking in making agroforestry an instrument for transforming the lives of the rural farming population, protecting ecosystem and ensuring food security through sustainable means. The major highlights of the Policy are: establishment of institutional set-up at the national level to promote agroforestry under the mandate of the Ministry of Agriculture, GoI; simplify regulations related to harvesting, felling and transportation of trees grown on farmlands; ensuring security of land tenure and creating a sound base of land records and data for developing an market information system (MIS) for agroforestry; investing in research, extension and capacity building and related services; access to quality planting material; institutional credit and insurance cover to agroforestry practitioners; increased participation of industries dealing with agroforestry produce, and strengthening marketing information system for tree products. Initially 20 important multipurpose tree species have been identified at the national level to be exempted from all restrictions related to harvesting, transportation and marketing grown under agroforestry systems. One of the objectives of NAP, 2014 is to bring together various programmes, schemes, missions among the elements of agroforestry under one platform functioning in various departments of agriculture, forestry and rural sectors of the government. It is proposed to be achieved through setting up of a National Agroforestry Mission/ Board under the Department of Agriculture and Co-operation (DAC), Ministry of Agriculture, GoI and upgrading of NRCAF, Jhansi (now CAFRI, Jhansi) as a nodal centre with agro-ecology-based regional centres in different parts of the country. This step will promote value chain, climate-resilient technology development and pave the way for region-based marketing linkages in agroforestry. The Policy also suggests massive extension programmes in order to broadcast the outcomes of intensive R&D activities in the field of agroforestry.

## Significance of Agroforestry

Agroforestry provides a different land use option, compared with traditional arable and forestry systems. It makes use of the complimentarily relationship between trees and crops, so that the available resources can be effectively utilized. It is a practice that supports the environment and has an obvious landscape benefit (Boyd, 2007). Efficient, modern versions of agroforestry have been developed, that are adapted to the constraints imposed by mechanization. The agroforestry plot remains productive for the farmer and generates continuous revenue, which is not feasible in arable land (Westoby, 1975). Agroforestry allows for the diversification of farm activities and makes better use of environmental resources. Owing to increase in population of human and cattle, there is increasing demand of food as well as fodder, particularly in developing countries like India. Each year farmers of the world must now attempt to feed 81 million more people irrespective of weather. It is important to note that there will be 19% decline in cropland per head by the end of this century due to population explosion. Therefore, there is slight scope to increase food production by increasing the area under cultivation (Bene, 1977). A management system therefore, needs to be devised that is capable of producing food from marginal agricultural land and is also capable of maintaining and improving quality of producing environment. Agroforestry has interesting advantages from three different perspectives.

**From the agriculture perspective**

1. Diversification of the activities of arable farmers, with the building-up of an inheritance of multi-purpose trees, with continuous revenue from farm.
2. Protection of intercrops and animals by the trees, which have a windbreak effect, providing shelter from the sun, rain, wind, soil erosion and stimulating soil microfauna and microflora.
3. Recycling of some of the leached or drained nutrients by the deep roots of the trees; enrichment of the soil organic matter by tree litter and by the residue of the trees.
4. Possibility of combining the interest of the farmer (for an inheritance of wood) and the farm (for access to cultivated land). Possible increased remuneration for the arable farmer for the trees.
5. An alternative to full reforestation of arable land, permitting the continuation of arable activity on land whose arable potential otherwise is conserved.
6. In silvipastoral plots, fodder units can be available at different periods compared to full cropped plots, extending the grazing calender.

**From the forestry perspective**

1. Enhancement in the biomass growth of the trees by wide spacing (+80% over 6 years in the majority of the experimental plantations). A large reduction in the maintenance costs of the plantation, due to the presence of the intercrops.
2. Improvement in the quality of wood produced (wide regular rings, suited to the needs of industry), because the trees are not subjected to cycles of competition and thinning.

3. Guaranteed follow-up and tree care due to the arable intercropping activity. In particular, protection against the risk of fire in susceptible areas, with pasture or with intercrops like vine or winter cereals (clear bare ground in summer after stubble ploughing).
4. Agroforestry plantations on arable land allow the development of a quality wood resource that complements rather than competes with the products from traditionally exploited forests.

### From the environmental perspective

1. Improvement to the development of natural resources: the total wood and arable production from an agroforestry plot is greater than the separate production obtained by an arable-forest separate cropping pattern on the same area of land.
2. Better control of cultivated areas of land: by substituting for arable plots, the agroforestry plots contribute to diminishing the cultivated area of land.
3. Creation of original landscapes that are attractive, open and favor recreational activities. Agroforestry plots have a truly innovative landscaping potential, and would improve the public image of farmers to society.
4. Counteract the greenhouse effect: constitution of an effective system for carbon sequestration, by combining the maintenance of the stock of organic material in the soil (the case especially with meadows), and the superimposition of a net fixing wooded layer.
5. Protection of soil and water particularly in sensitive areas.
6. Improvement of biodiversity, especially by the abundance of "edge effects". This in particular, permits a synergistic improved integrated protection of crops by their association with trees, chosen to stimulate the hyperparasite (parasites of parasites) population of crops, is a promising way forwards.

## Future of Agroforestry

The start of organized research in agroforestry in early 1970s has taken different dimensions. Earlier, agroforestry was considered to provide bonafide needs of the communities by incorporating trees, crops and animal component in unit land through varying interfaces. The main emphasis was on tangible out-puts like biomass from trees and crop yield. But, the current understanding of agroforestry systems is changed and now agroforestry is considered a problem-solving science (Dagar and Tewari 2016). It is considered as a better climate change mitigation option than ocean and other terrestrial ecosystems because agroforestry provides ecosystem services such as carbon sequestration, biodiversity conservation, water quality enhancement and land restoration. However, lack of predictive metrics and the regulatory environment mechanisms are impeding the adoption of several ecosystem services. The major issues of maintaining food and fiber production, soil degradation, biodiversity decline and climate change mitigation can be managed at a meaningful scale with sustenance through agroforestry option. Industrial agroforestry involved short rotation forestry to accommodate agricultural crops in fast growing plantations. The impact of agroforestry on the environment occurs at a range of spatial and temporal scales from fine scale impacts on soil structure and quality to impacts on the society at regional and global scales. Future $CO_2$ mitigation initiatives will greatly benefit from incorporation of biodiversity component into the design, implementation and regulatory framework. This will also bring the two quite different entities of biodiversity conservation and carbon sequestration together. The convergence of carbon sequestration with biodiversity conservation and salinity management presents a unique opportunity to tackle environmental problems together. Using an agroforestry approach will allow the integration of trees into farmland without displacing food production, but rather stabilize agricultural systems.

It was included in global program such as Reduced Emission from Forest Degradation and Deforestation ($REDD^+$) related to climate change mitigation and adaptation. The gains and developments of last two decades of agroforestry research are certainly impressive. The question, then, is what next? What direction should agroforestry research take? In order to answer this, we need to critically evaluate: (1) the extent to which the original research/development expectations and goals – whether stated directly or implicitly two decades ago – have been fulfilled, and (2) the current and emerging problems and expectations. Although the terms 'research' and 'science' are often used together, they are neither inseparable nor not synonymous. It is true that research is done according to scientific procedures, but research does not, and need not, always result in advancement of basic science. According to dictionary definitions, science is systematized knowledge derived from observation, study, and experimentation carried on in order to determine the nature or principle of what is being studied; it refers to the state of fact or knowledge, as opposed to intuition or belief. Research is careful, patient, systematic, diligent inquiry or examination in some field of knowledge, undertaken to establish facts or principles, and involves laborious or continued search for truth. Thus, while all research must be carried out scientifically, the objective of the research could be the advancement of knowledge (science) or the application of knowledge; the latter aspect, which involves the study of the practical or industrial application, is usually referred to as technology.

Development of appropriate biological technologies forms only one, perhaps the first step in the promotion of agroforestry; provision of public policies and institutions that support program implementation is an equally important requirement. Analysis of the experience of several countries have shown that rapid and dramatic expansion of agriculture and plantation forestry occurred when supportive public policy was put in place through appropriate subsidies, national investment, and adequate extension programs. Carbon sequestration in terrestrial ecosystems is a good example of such larger-scale spatial issues. Agroforestry systems such as fuelwood plantations, shelterbelt/windbreak systems and woodlots may have the potential to sequester C, or offset fossil fuel emissions by substituting sustainably produced fuelwood and fodder. Based on a preliminary assessment of national and global terrestrial C sinks, it is argued that there are two primary beneficial attributes of agroforestry systems in this regard: (1) direct near-term C storage (decades to centuries) in trees and soils and (2) potential to offset immediate greenhouse gas emissions associated with deforestation and subsequent shifting cultivation. The impact of agroforestry on terrestrial C budget is an area that needs to be researched. Other environmental benefits such as water-quality improvement through agroforestry have received no attention in through research in the tropics. Results from temperate-zone agroforestry point to the beneficial effects of riparian buffer strips in reducing sedimentation load and nutrient pollution in agricultural landscapes. Conservation of biodiversity is another issue in which agroforestry may have a major role to play but which has received scant attention. It is obvious that agroforestry systems are far more diverse than crop or tree monocultures. Furthermore, the most important way in which agroforestry can help conserve plant and animal biodiversity lies in reducing tropical deforestation. It may not be prudent, however, to expect that having a high number of economically useful plants per hectare would lead to biodiversity conservation. Similarly, in commercial prospecting among natural sources for new products (such as the biodiversity prospecting, which is usually promoted not because of the realization of the value of nature's diversity for ecological, aesthetic, or spiritual reasons, but by the lure for the discovery of the miracle plant or animal that contains the cure for such human diseases as cancer or AIDS), the value of biodiversity is not as high as some conservationists might suppose. Thus, aroforestry teaching, research and extension should involve building bridges *from* the past *through* the present *to* the future. During the past two decades, we have progressed through the stages of inductive reasoning and formulation of hypotheses to deductive experimentation. We should now place greater emphasis on harnessing the *science* of agroforestry to develop technologies of wide applicability. It is not a question of choosing either the advancement of science or the development of technologies; one cannot proceed without the other. However, we should not lose sight of the fact that agroforestry research was initiated to help solve problems. Therefore, the initiatives should be how to blend science and technology in an optimum balance in order to develop science-based technologies that will address the problems. Future generations will judge the merits of our efforts based on our success in accomplishing this important task of balancing science and technology.

## Conclusion

The plant husbandry happened to start under progressively arid climatic zone from about pre-Neolithic period in India. The role of many common trees such as khejri or sami (*Prosopis cineraria*), aswattha (*Ficus religiosa*), palas (*Butea monosperma*) and varana (*Crataeva roxburghii*) in Indian folk-life has been mentioned in ancient literature of Rig Veda, Atharva Veda and other ancient scriptures. The agroforestry practices believed to be originated during Vedic era in 1000 BC but introduced as science only in recent 3 to 5 deacdes. The systematic research in agroforestry geared up after the establishment of the International Council for Research in Agroforestry (lCRAF) in 1977. In India, the systematic work on agroforestry was initiated during the 7$^{th}$ Five Year Plan i.e. in late 1970s. It is further geared up with the inception of AICRPF by the ICAR and degree program in forestry/agroforestry by the state agricultural universities. The role of agroforestry in meeting present or future requirements of tangible and non-tangible benefits has been very well recognized. Traditional agroforestry systems manifest rural people's knowledge and methods to benefit from complimentary uses of annuals and woody perennials on the sustained basis. Shifting cultivation, taungya, homestead gardens, scattered trees on farm lands, woodlots, shelterbelts, trees on rangelnads are some of the main traditional agroforestry systems in practice. Agroforestry has not gained the desired importance as a resource developmental tool due to its dependency on multi-institutions and multi-disciplinary approaches. Even, there is serious lack of institutional mechanisms to promote agroforestry at all levels. Therefore, India has announced the landmark National Agroforestry Policy 2014 that will mainstream the growing of trees on farms to meet a wide range of developmental and environmental goals with addressal of all issues of agroforestry. There is need to develop location-specific, need-oriented systems along with the necessary support systems so that farmers can get the required inputs easily and market the produce at competitive prices. The exploitative tendency of modern agriculture, aimed at higher and higher production per unit area, is in fact rapidly degrading the basic production capacity of the ecosystem. There is aberrations in the estimates of area under agroforestry by ICAR-CAFRI, Jhansi and Bhuvan LISS III. The accuracy in the area can be achieved by developing digital signatures for all the important agroforestry tree species. There is increasing demand of food as well as fodder, particularly in developing countries like India owing to increase in human and cattle population. Each year farmers of the world must attempt to feed 81 million more people irrespective of weather. It is important to note that there will be 19% decline in cropland per head by the end of this century due to population explosion. Therefore, there is slight scope to increase food production by increasing the area under cultivation. A management system therefore, needs to be devised that is capable of producing food from marginal agricultural land and is also capable of maintaining and improving quality of producing environment. Agroforestry has interesting advantages from agriculture, forestry and environmental perspectives. The current understanding of agroforestry systems is changed and now agroforestry is considered a problem-solving science. It is considered as a better climate change mitigation option than ocean and other terrestrial ecosystems because agroforestry provides ecosystem services such as carbon sequestration, biodiversity conservation, water quality enhancement and land restoration. There is plethora of active markets of ecosystem services obtained from the developed agroforestry systems. The emphasis should be on harnessing the *science* of agroforestry to develop technologies of wide applicability. Therefore, the future initiatives should be on how to blend science and technology in an optimum to balance and develop science-based technologies that will address the present problems and future challenges. Future generations will definitely judge the merits of the efforts based on the success in accomplishing the gigantic task to use the natural resources on sustainable basis.

# Chapter - 3

# Planning and Development of Agroforestry Practices

*Aman Dabral, Prshanth G. Naik., Poonam, Dhanush C. Reddy*

## Introduction

In the present scenario population growth and its related issues are the major problems facing by many tropical and subtropical nations. Rapid growth of population exerts heavy pressure on land, water and other natural resources (FAO, 2011, Shaw 1992, Alston *et al.*, 1995.). Suitability of agriculture land, grazing land, and forestry uses for people has been reduced because of agricultural and forest products from economically developed and developing nations increasing pressure and the demand. This shortage of land as well as land productivity has resulted in the over use of land and in deforestation, which in turn contributes to soil erosion and related environmental degradation. Ecologically overpopulation may leads shortage of living place, food, fresh air, water and depletion of nature and its resources on other hand economically problems arise due to overpopulation can be poverty, unemployment, low per capita income, reduced savings and investment, trade deficit, growing indebtedness, increasing welfare expenditures, unemployment. With respect to the all mentioned negative consequences, agroforestry has been suggested as one means of ameliorating negative land-use impacts while addressing the basic needs of agrarian societies. (Duchhart *et al.*, 1989).

In a broad science agroforestry is a practice of growing trees on farm land with or without in combination of animals for creating more profitable, productive, protective, ameliorative and sustainable land use systems (Nair, 1987.,Alao, J.S. and Shuaibu, R.B. 2013.,Fratini *et al.*, 2016., Tiwari *et al.*, 2017). Major components of the agroforestry practices include trees, agriculture crops and animals (Nair, 1987). Agroforestry objectives need a step by step procedure to finally achieve the desired goal and this step constitutes the agroforestry planning. Agroforestry planning is an act or process of organizing the strategies concerned with the policy formulation related to management of agoforestry components required to achieve a desired goal. The development of agroforestry planning is much useful for the supervision of the agroforestry practices. It also maximise the chances of getting success in the proposed policies. Even time also have maximum importance in the process itself and it can be properly utilized using a proper agroforestry plan.

## 1. Agroforestry System Necessity is there or not?

Before initiating the agroforestry planning process it is compulsory to know about the necessity of particular agroforestry system in specific region is there or not. Step wise procedure of the initial process is as follows;

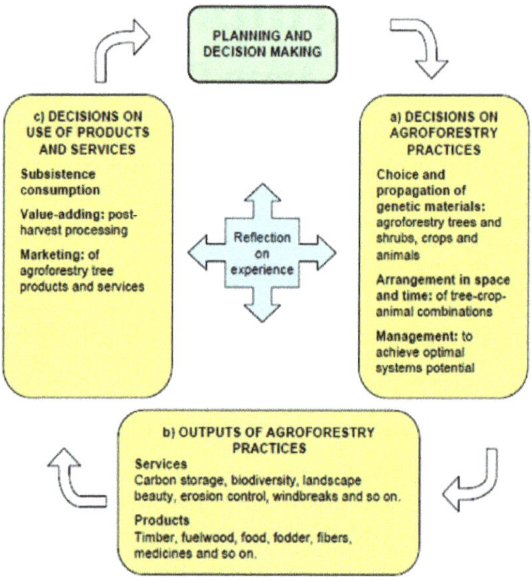


**Figure 7. A model of farmers' agroforestry production cycle**

i). Description of farmer and community requirements. Farmer requirements include continuous income availability and also higher profit on the investment. Community requirements include continuous employment, protection benefits to the community properties.

ii). List down all the needs that can be fulfilled using the agroforestry design. After describing all the requirements it is necessary to cross check those with all the needs that can be achievable using the agroforestry design.

iii). List down all the future prospects of an agroforestry system already present in the locality in question. These step focuses on finding out all the future benefits that can be possible with the existing agroforestry design in all levels of economical, environmental, social etc.

iv). Find out the limiting factors present in agricultural system. Limiting constraints includes.

v). Considering the capability of the people to adopt the agroforestry system.

vi). Finally conclude weather it is worth to develop the system or not.

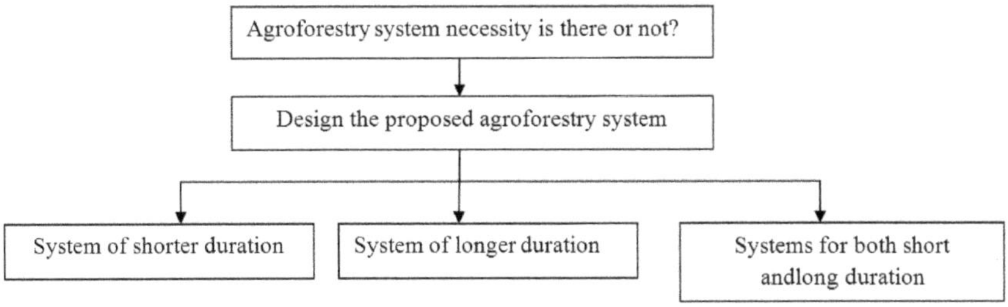


## 2. Design the Proposed Agroforestry System

After knowing necessity of agroforestry system for particular area by taking above mentioned requirement, future prospects, limiting factor and people capability next step is to design appropriate system which include following steps.

i). Selection of area.

ii). Suitability of existing soil, water and crops with respect to the design.

iii). Selection of species combinations (trees, shrubs and grasses to be used).

iv). Characterization of minimum requirements (space, water, fertilizers and shade tolerance).

## 3. System of Shorter Duration

i). Make the strategies for the control of soil erosion, gully fixing and earth operations.

ii). Plan the plan the spacing of trees and shrubs on the land as per the final space requirements.

iii). Plan a succession of annuals crops. Sometimes even the short lived perennials are also preferable. Shade tolerance is the most important criteria one has to consider before selecting the crop.

iv). Then select the crops which are suitable in the final year of the intercropping. In a broad way intercropping denotes to growing the agricultural crops in between the rows of the forest tree species.

## 4. Systems of Longer Duration

More care should be taken while planning of the systems of longer duration. As a name indicates this systems lasts for long time. So multiple benefits are advisable to consider while planning a design. Sustainability is the primary focus of the planning of long duration agroforestry designs.

i). Maintaining the proper proportion of agroforestry component is the most important step in the long run. So the proper proportion of fruit trees and lumber trees should be decided by analysing the preference of the farmer.

ii). Similarly spacing of the trees should be made by considering the final space requirement of the tree crops.

iii). Succession of the annual and understory crops are important factors to consider. Crops which have protection functions and soil enrichment properties are need to get the much needed preference.

iv). When the trees are getting the large structure it is necessary to make the adjustments for the planting plan of the most shade tolerant crops. These shade tolerant crops are grown on the shadiest places created by the large trees.

## 5. Systems of Both Temporary and Permanent Nature

Planning in this case is the most important thing as it needs to focus on both temporary and permanent needs. Sustainable benefits are the primary focus of these systems.

i). In an agroforestry practices protecting the land is the most important task. Bare lands get unnecessary attention from the various problems. So ground vegetation is recommended to cover the bare land in order to protect the soil from scorching effect of sun and also erosion.

ii). To check about the effects of the implementing designs, it is always preferable to go for a small scale planning at the beginning stage.

iii). Inputs and outputs are also the most important factors of agroforestry planning. It is a universal fact that one always wants more output compared to the inputs he is going to offer. So measurements of the inputs and outputs are the primary important thing to do.

iv). Ultimate success of the planning is most important and major objective of any agroforestry system. So the final evaluation of planning for the confirmation of weather the final objectives is met or not is the most important task to accomplish.

Before implementing the new system it is advisable to know about the effects of the introducing system on the environment in general and also the existing system practicing. So cautions should be taken to avoid any further complications in the further stages.

### Advantage

1. Planning is responsible for more efficient and quick achievements in the agroforestry practices.
2. Planning also gives additional power to the already existing agroforestry designs for its continuous growth.
3. Agroforestry planning provides a set of instructions by focusing on the objectives which is helpful for avoiding the unintentional repetations in the work.
4. It minimizes the cost of agroforestry operations.

5. Unidirectional flow of work is possible without any deviations from the main objectives.
6. It provides control over the components behaviour involving in agroforestry practices.
7. It facilitates the process of decision-making.
8. It enables the management to decide the future programmes by analysing all the resources which are present and also by considering the past performances.
9. It helps in adaptation of community with the changes in technological development.
10. It avoids waste of men, money, materials and machinery. Thus planning indirectly leads to large-scale economies.
11. Planning gives direction to the efforts of the organization to useful purpose, rather than being wasted.

### Disadvantages or Limitations of Planning

1. Predicting the future is a hypothetical process. So accuracy of informations about the future is the constraint of the agroforestry planning.
2. Time, money and effort are required in the formulation and revision of plans. It is a time consuming process. It is an expensive process. Usually cost of plans making is more than the actual profit it generates.
3. Analysis of all resources takes much time. Hence taking quick decision is not possible even though it is the most important task to do.
4. Planning gives a false hope like implementation of plan will be useful to solve all problems. But in reality there is a chance of failures of plans due to several constraints.
5. Repetitive operations are must and should to keep the significance of the executed plan.

## Complex Process of Agroforestry Planning According to Gold et al, (2013)

Developing a complex plan may seem tough task at the beginning, but the long-term benefits far outweigh the difficulties of plan cre¬ation. By such having a plan, farmer can better know how to plan an agroforestry practice to their farm. And, the planning process also will help develop a familiarity with the management required to reach the goals, objectives, benefits and economics desired. The best way to ensure success is by thought-ful and honest planning.

## Steps for developing a complex Agroforestry Plan

### Personal Assessment

***Step 1: Initial Objectives and 1Priorities***

Land management objectives and priorities will be specific to the circumstances and the area you want to develop for agroforestry. However, among other possibilities, you may want to:

- ❑ Develop a new source of income from unproductive land (diversification).
- ❑ Reduce costs of an existing farm or forest operation.
- ❑ Develop a source of long-term income.
- ❑ Develop a source of short-term income while awaiting long term income from timber.
- ❑ Reduce property taxes.
- ❑ Protect or improve environmental conditions.
- ❑ Increase grazing opportunities.

***Step 2: Evaluate Personal Resources***

An evaluation of the resources available to you will help determine which agroforestry options are best suited to your operation. An evaluation should include:

- ❑ Management and labor availability
- ❑ Equipment and buildings
- ❑ Livestock.
- ❑ Plant material.
- ❑ Other materials – Resources such as sawdust or shavings, manure and straw for mulch.

### Biophysical Site Assessment

***Step 3: Identify Current Land Uses***

List the present uses of each area of your land and the products you harvest, and record them in Workbook. These uses could include:

- ❑ Residential
- ❑ Recreational
- ❑ Farming (which crops)
- ❑ Grazing (type of livestock)
- ❑ Timber production
- ❑ Non-timber production
- ❑ Environmental use

### *Step 4: Map Area(s) for Agroforestry Development*

An agroforestry development may include all of your land or only specific areas, such as existing woodlands, open field, loggedover area or riparian zone. In either case, identifying objectives and making decisions will be much easier if you break your land into separate development areas with similar current uses and site conditions (such as soil, moisture and existing plants).

### *Step 5: Climate Assessment*

Climate on your site ultimately determines what you can grow on your land. It will establish the range of possibilities for your agroforestry practice

### *Step 6: Soil Assessment*

Assessment of soils properties should be done taking into consideration following points.

- ❑ Land Capability Classification
- ❑ Soil Texture and Composition
- ❑ Soil Depth
- ❑ Soil Moisture
- ❑ High Water Table, Standing Water or Flooding
- ❑ Nutrients

### *Step 7: Physical Features (Terrain)*

There are several physical features that can influence your capability to produce particular crops on a site.

- ❑ Aspect
- ❑ Terrain relief
- ❑ Altitude

### *Step 8: Timber and Non-Timber Forest Crop Inventory*

Tree inventory: The tree inventory described here is only intended to determine the non-timber resources available (e.g., boughs, vines, bark, burls).

- ❑ Number of trees and species
- ❑ Tree age
- ❑ Tree height
- ❑ Condition of trees
- ❑ Crown closure

Non-timber Inventory: For each species, record the following:

- ❑ Total number of plants for each species
- ❑ Estimate of the percentage cover for each species
- ❑ Proportion of harvestable vs. non-harvestable individuals

## Agroforestry Development Ideas

### *Step 9: Agroforestry Ideas – Brainstorming*

Brainstorming is the process of creating ideas in mind and writing them on paper. List all the ideas and try to prioterize the ideas. You can reduce this list to 'best bets' later. Input from all the family members and people can help generate useful ideas. This will lead to generating more and more ideas through various suggestions. For each development area, sort the list by practice and record this information in the corresponding Workbook form. In general, more intensive agroforestry practices (e.g., forest farming) are suited for smaller units of land, while the more extensive practices (e.g., silvopasture) are suited to larger units of land.

### *Step 10: Listing 'Best Bets. The best plan should be clearly defined for success of agroforestry model.*

### Evaluate the 'Best Bets' in the Context of the Industry

The intent of this short but important step is to create a list of potential crops that can be grown on your land. This list of 'best bets' will be based on the information you have gathered so far – including land and personal resources, site conditions and current land use – combined with the plant information provided.

### *Step 11: SWOT Analysis*

Having analyzed the current situation of your farm with respect to objectives and priorities, personal resources, site assessment and vegetation inventory, and brainstormed ideas of agroforestry practices, you can use that information to identify Strengths and Weaknesses, Opportunities and Threats or a SWOT analysis for the crops you think may produce. This can be helpful in defining and clarifying the issues you need to address in the rest of the planning process.

### *Step 12: Porter Five Forces Model*

The Five Forces Model serves as a good framework for assessing different industries you would like to get in. The Five Forces Model identifies coordination and control aspects of an industry and provides a guideline for understanding the resources and relationships needed to be successful in a market.

### *Step 13: Revising Your 'Best Bets'*

In Step 11 of your agroforestry plan, you identified the plants that can grow on your land, and which you can produce with the resources you have available. After a thorough analysis using the SWOT Analysis and Porter Five Forces Model, the list of 'best bets' can be refined to include those plants and products you think have the best market potential.

## Marketing Strategy for 'Best Bets'

### *Step 14: Select and Describe Target Market(s)*

Your first task in building a customer strategy is to identify your target market. Target markets are most commonly characterized as either individual households or businesses. Begin your target market research by developing a customer profile. Customer profiles can help you determine if a market segment is large enough to be profitable.

### *Step 15: Adding Value to Products*

As you think about the products your business will offer, try to describe them in terms of the value they will bring to your customers. List all value-added opportunities and identify 'pros' (how it will benefit the needs of each customer segment) and 'cons' (e.g., costs, risks) for each of them. Identify also unique characteristics that will differentiate your product from competitors. Define each product to address specific needs for each market segment.

### *Step 16: Getting Products to the Buyer*

Now that you have a customer and product in mind, your next task is to identify how to move or distribute products from your farm to the customer's house or store shelves.

### *Step 17: Setting the Price*

In general, pricing strategies are based on two factors: Prevailing market prices and your costs. In the long run, your price has to cover your full costs – including production, marketing and promotion – as well as a return for your time and investment.

### *Step 18: Promoting Your Products*

Promotion is very important to gain product recognition among customers. Promotional strategies often are built around a "message." The message that you deliver about your product or business is just as important as the product itself.

## Agroforestry Practice Design and Management

### *Step 19: Revisit Your Objectives and Priorities*

The first step in developing your agroforestry plan was to outline your initial objectives and priorities for your land-based business. Since then, you have gathered considerable amounts of information that will likely change at least some of those original goals.

### *Step 20: Detailed 'Best Bets' Crop Information*

To develop the most functional and profitable agroforestry practices, you should know as much as possible about each plant you intend to grow. You will have gathered much of this information while filling out your Workbook, using your land assessment, plant inventory (where appropriate), market research, Appendices 2 and 3 of the Training Manual, and possibly your own additional research.

### *Step 21: Designing Your Agroforestry Practices*

Now is the time to arrange your plants into agroforestry practices that best meet the management objectives you outlined in Step 19. The information on each plant crop, and the management information provided earlier in this chapter, will allow you to design and manage agroforestry practices that best meet your objectives.

### The Agroforestry Development Plan

The purpose of building an agroforestry development plan is to create a schedule of the work that needs to be done in the years ahead to fully develop your agroforestry area(s). Your completed Agroforestry Development Plan will include specific tasks, timelines and labor projections for each agroforestry practice on a year-by-year basis.

#### *Step 22: A Five-Year Management Projection*

Your five-year management projection is a schedule of the work you plan to undertake in the next five years to develop your agroforestry practices. The specific tasks and timelines you prepare will form the basis for your yearly activity schedule

#### *Step 23: Yearly Activity Schedule*

The yearly activity schedule describes specific tasks that need to be done, when and by whom. This is the document you will use to plan your work schedule on a month-by-month basis. A good yearly activity schedule will allow you to identify potential time and business conflicts, and ensure you avoid overlapping seasonal activities that could create shortages of labour and resources.

## Scope of Agroforestry planning

Each plan requires careful balance between all components, practices, processes, tools and techniques etc. to ensure that a particular plan is appropriate and also it makes sure of getting success. Thus agroforestry planning provides a framework of operation that is going to be carried out in a land. An agroforestry plan involves developing a vision and identifying strategies that address current and future issues to ensure the long term prosperity and resilience. In future, pressure on land and forest will be increasing due to demands for their produce. So appropriate agroforestry planning will be helpful to nullifying that pressure in future.

## Conclusion

Basically planning programme gives an idea about what to do in future to get higher profits and also it provides a strategy for utilizing the resources already existing. It also provides a framework of methods by following that we can achieve desired goals. Planning process involves knowing about the necessity of planning at the initial stage of the plan making. Appropriate design for particular object should be formulated for shorter, longer and sometime in combination of both. In spite of the serious limitations, planning is still recognized as the basic requirement of the sustainable use of agroforestry. It is no doubt that unplanned operations will produce chaos and disorder everywhere without exception.

# Chapter - 4

# Designing Field Experiments in Agro-forestry

*Raj Kumar, Bharti*

Designing of experiment means planning an experiment for assigning treatments (various objects of comparison in a comparative experiment) under investigation to experimental units in such a way that nuisance variables controlled as much as possible. Basically, it includes planning of the experiments, obtaining important information from it about the statistical hypothesis under study and data analysis. There are mainly three phases of every assignment i) Planning Phase which includes hypothesis building, choice of response variable, selection of factors and their levels etc. ii) Design Phase which includes number of observations to be considered, randomization technique to be used, mathematical model to describe the experiment etc. iii) Analysis Phase which includes data collection, tabulation, analysis and interpretation of data.

For the validity of statistical analysis, three basic principles i) replication ii) randomization and iii) local control are observed according to Prof RA Fisher who pioneered the study of experimental design. Replication is the repetition of treatments by applying them to more than one experimental unit. It is necessary to replicate treatment to get an estimate of experimental units. Replication along with randomization give the valid estimate of error and along with local control diminishes the error. Randomization is essential for valid estimate of experimental error and it also minimizes the biasness of the results. This principle is essential keeping in view the assumption of analysis of variance techniques regarding randomness of sample. Randomization by itself is not sufficient it must be supported by repetition of treatments for the validity of the experiment. The principle of local control is useful only when the experimental material is heterogeneous. We have to divide the heterogeneous field into a number of blocks and blocks are formed in such a way that there is minimum possible variation within block and maximum possible variation among blocks.

Agroforestry is a collective name for land - use systems and technologies where woody perennials are intentionally grown on the same land management units along with agricultural crops and/or animals in some form of spatial arrangement or temporal sequence. These systems are more complex than the mono-cropping systems. In addition to the basic principles i.e. randomization, replication and local control, there are quite a lot of other factors that have to be taken care of while planning agroforestry experiments.

## RBD

The most widely used design because of its simplicity is Randomized Complete Block Design (RCBD). In field experiments, if the whole of experimental area is not homogeneous and the fertility gradient is only one direction then a simple method of controlling the variability of the experimental material consists in grouping the whole area into homogeneous subgroups. These sub groups are called blocks. These blocks are made perpendicular to the direction of the fertility gradient. If the treatments are applied at random to relatively homogeneous blocks, then the design is RCBD. For example, Randomized Complete Block Design was used to identify suitable crop among five crops (moong, moth, cowpea, guar and bajra) under different tree species (*Acacia nilotica, Acacia tortilis and Dalbergia sissoo*) at Fatehpur Shekhawati during 1986. This design is not suitable for large number of treatments (greater than 20) because in that case the block size will increase and as such will not remain homogeneous. Sometime it is also desirable to use Factorial RCBD in agroforestry, especially when our objective is to test means of two or more than two factors with two or more than two levels of all factors simultaneously in an experiment.

## Split Plot Design

Another commonly used design is split – plot design. In field experiments, the efficiency of RCBD is found to deteriorate when number of plots per block is large. Similar in case of factorial experiments when treatment combinations become large and heterogeneity is not controlled, confounding helps to some extent but suffers from the drawback of scarifying the information regarding higher order interaction. In such cases, split plot design may be adapted. For example, in an alley cropping (planting of trees or shrubs in two or more sets of single or multiple rows with agronomic, horticultural, or forage crop cultivated in the alleys between the rows of woody plants) experiment under the component and system management trials, for testing hedgerow species and hedgerow pruning management, species could be allocated to main plots and pruning management (frequency, height of cutting, etc.) to subplots of a split-plot design. But these experiments should be carefully used because the treatments in the main – plots are compared with less precision than those in the subplots. Split block design is analogous to split plot arrangement is recommended for two factor experiments when both factors require large plot sizes and when the interaction between the two factors is desired to be measured more precisely than the effects of either of the two factors.

RCBD, Split Plot Design, Split Block Design are randomized designs. But sometimes randomization is impracticable. For example in spacing trials, the extreme treatments must not adjoin. The designs for such situations are called systematic designs and these can be used in agroforestry experiments (Huxley, 1985). Many factors (slope, rainfall, distance from tree etc.) in agroforestry research do not allow randomization. These designs despite their limitations have been proposed for agroforestry research because they minimize the requirement for field and experimental resources.

Augmented designs is a very powerful and appropriate approach for agroforestry research. The technology developed at research stations is tested under farm conditions to obtain information about its performance and acceptability by the farmers. Selection of treatments and choice of experimental design is an important factor in on-farm trails. The experiments here use fewer, but larger plots and plot choices are limited. To meet such situations augmented designs can be adopted.

## Agroforestry Diagnosis and Design

Agroforestry D & D is a family of procedures for the diagnosis of land management problems and potentials and the design of agroforestry solutions. The ICRAF has developed an approach to assist agroforestry researchers and development fieldworkers to plan and implement effective research and development projects.

## Key Features of D & D

**i) *Flexibility:*** D & D is a flexible discovery of procedure, which can be adopted to fit the needs and resources of different users

**ii) *Speed:*** D & D has been designed with the option of a 'rapid appraisal' application at the planning stage of a project with In-depth follow-up during project Implementation.

**iii) *Repetition:*** D & D is an open-ended learning process. Since initial designs can almost always be improved. The D & D process need not end until further Improvements are no longer necessary.

### Basic logic of AF Diagnosis and Design

| Basic question | Key factors to consider |
|---|---|
| **Prediagnostic stage**<br>Which land-use system?<br>How does the system work? | Distinctive combinations of resource technology and land-user objectives.<br>Production objectives and strategy subsystems and components. |
| **Diagnostic stage**<br>How well does the system work? | Problems in meeting objectives, causal factors, constraints, and intervention points. |
| *Design and evaluation stage*<br>How can the system be improved? | Specifications for problem-solving or performance-enhancing interventions. |
| *Planning stage*<br>How can the Improved technology be developed and disseminated? | D & D needs, extension needs. |
| Implementation stage<br>How can the plan of action be adjusted to new information? | Feedback from research trials, independent farmer innovations etc. |

## Procedures of AF Diagnosis and Design

The procedures of AF D & D are usually done of two types:

1) 'Macro' D & D and 2) 'Micro D & D '

### i) MACRO D & D

An agroforestry research programme normally begins with a macro D & D exercise covering an entire ecological zone within a country. It consists of a rapid appraisal, primarily based on secondary information complemented by a few required surveys in the field.

Macro D & D consists of assessment of existing land use system problems, agricultural policies and institutional arrangement, current agroforestry practices and the potential agroforestry systems that can be developed for improving productivity and sustainability. The study area selected should be represented in such a way that its piutcome can be replicated at national levels. Study area is selected usually based on the following factors.

- Contribution to food production and the national economy;
- Population area
- Urgency of problems or importance of unexploited potential
- Level of agricultural development and land use intensification

### ii) MICROD & D

A main aspect of macro D & D is the selection of land-use systems within the chosen ecological zone. This leads to the selection of a particular system for more detailed study by micro D & Ds. A land-use system is should be distinctive combination of crops/annuals, livestock, trees/perennials and other production components.

The primary aim of analysis of management unit in such a way that to make decisions and shares available resources, objectives, labour and products. In a land use systems all the characteristics should be analyzed that affect its management and performance. These characteristics include the following:

- Location: Administrative and political divisions
- Environmental characteristics:
- Socio-economic characteristics
- Land-use:
- Resources/supporting service
- Development activities and policies.

An important aspect of micro D & D is an analysis of the needs, output, objectives, and constraints of land-users. This step is based on interviews and field surveys.

One major aim of micro D & D is to analysis existing knowledge of agroforestry systems and practices. Type of trees or shrubs to be grown with what type of management procedures, with types of objectives and obtaining what type of yields? Such analysis helps in defining strategies for working with target land-users.

Initially a D & D exercise should focus on to the assessment the performance of existing systems and its role in meeting land-users' needs. is to assess how well the existing system is performing and meeting the needs of the land-users. Any improvement in the performance in systems can be evaluated by comparing existing resources and outputs of the farmers with the biophysical potentials in term of yields obtained from agroforestry experiments). The assessment must be able to distinguish distinguish between problems that can be solved and those that cannot. The main emphasis should be on the problems that can be addressed by adoption of agroforestry.

The potential interventions must be identified and analyzed in terms of their capacity to improve the identified constraints. Initially, not only agroforestry but all interventions are considered. For example, soil sodicity issue could be addressed by applying gypsum and chemical fertilizer to the plants. Each alternative is evaluated for its technical potential and feasibility in terms of resources and capabilities of the land users.

The research team for a micro D & D comprises 5 to10 biophysical and social scientists. This exercise is usually completed in about three months. However, the fieldwork component entails more contact with farmers, often including a formal survey of 50 to 100 individuals with a semi structured questionnaire. Each team may need to allocate considerable time to reviewing ongoing research and extension work in the selected land-use system.

Promising agroforestry technologies may be sketched and detail plan drawn on the basis of results from micro D & D, but a full evaluation requires information about technology performance under the specific conditions of the land-use system. If a technology is well known and some farmers have already adapted it successfully, then it can be recommended for further extension.

On the other hand, if the technology is new to the area and not well known or if it represents a major departure from the farmers' 'current practices, a research programme must be designed to test the components and management factors and make sure that the technology is well adapted to the target land-use system.

## Criteria of Good Agroforestry Design

A good agroforestry design should fulfill the following criteria:

i) Productivity: Agroforestry productivity can be improved with many ways viz., increased output from trees, enhanced yields of annual crops, reduction in crop/ tree inputs, efficient utilization of labour, diversification of production, satisfaction of basic needs and other measures of economic efficiency or achievement of biological potential.

ii) Sustainability: Agroforestry helps in maintaining sustainability of the production systems by conservation and management of the natural resources.

iii) Adoptability: Agroforestry systems should be acceptable by the farmers. This means that the technology has to fit the social as well as environmental characteristics of the land-use system for which it is designed.

## Design and analysis of agroforestry experiment

The basic design depends on the type of variable to be measured. These design, which seem to have the most potential for use in agroforestry are discussed under the following headings:

1. Simple design for investigating the tree/crop interface
2. Systematic design for spacing experiments;
3. General designs for agroforestry systems

## Simple designs for investigating the tree-crop interface

Generally two designs are used for investigating tree/crop interface:

i. Parallel rows of trees and crops, and
ii. Concentric rows of crops around a single tree

Figure 8 shows two designs given in Chetty (1986) which could be used to investigate the tree/ crop interface. The designs should be replicated unless they are purely observational and in the case of Figure 8a the orientation should be randomized (perhaps with just two possible orientations). In both cases the analysis will involve fitting response curves describing the change in the variable measured with increasing distance from the interface. In Figure 8a, the effect of orientation could be considered and in Figure 8b, aspect (north, south, east or west).

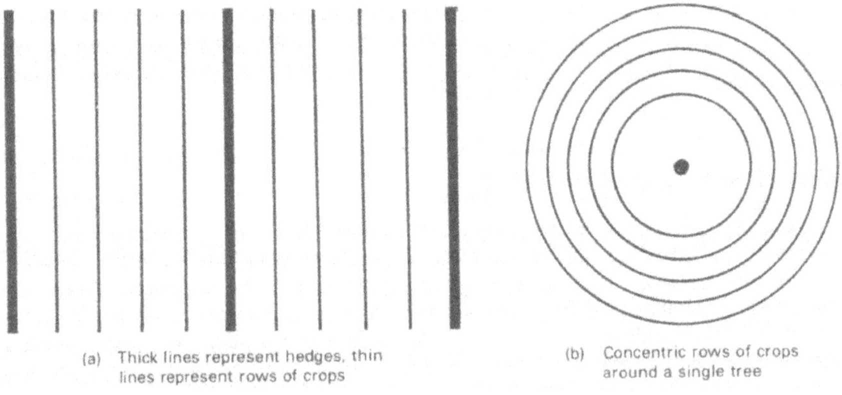


**Figure 8. Two designs for investigating the tree-crop interface: (a) parallel rows of trees and crops; (b) concentric rows of crops around a single tree. Thick lines represent hedges; thin lines, rows of crops.**

## Systematic designs

In some cases, randomization is not practically possible. The designs for such situations are called Systematic designs and these can be used in agroforestry experiments (Figure 9) (Huxley, 1985b).

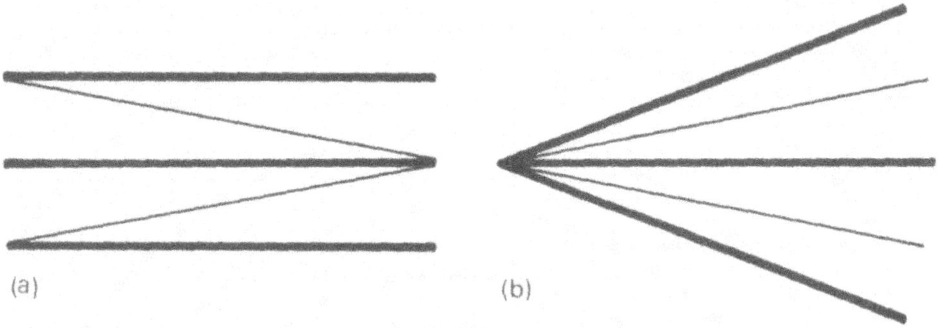


**Figure 9. Two very simple systematic designs**

Fan designs (Nelder, 1962) and Parallel row layouts (Bleasdale, 1967) are the systematic designs normally used to investigate sole crop spacing problems with a wide variety of plant material. For agroforestry containing tree-crop experiments, including tree-crop interaction studies, spacing is an important variable and systematic design can be designed to observe the effects (Figure 10).

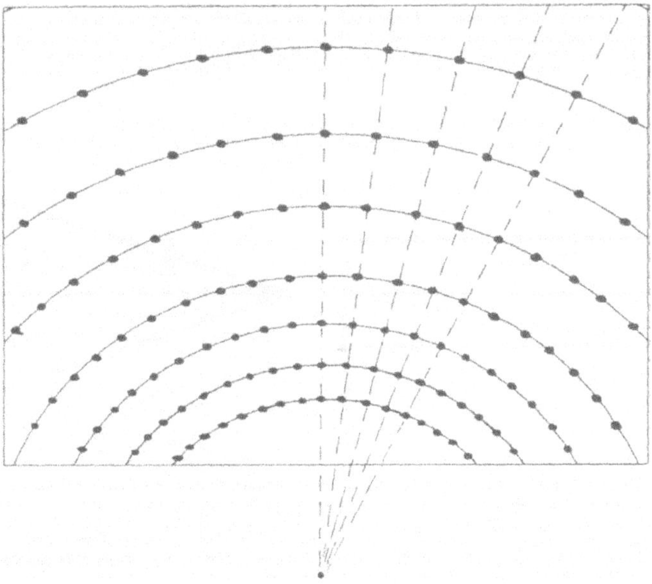

**Figure 10. A fan design (Nelder 1962).**

**A parallel-row design**: the previous designs suffer from the problems that the crops are not rectangular arranged, but lies on the area of the circle. However, paralleled row design avoids this although at the cost of confounding the effects of area and shape (Figure 11).

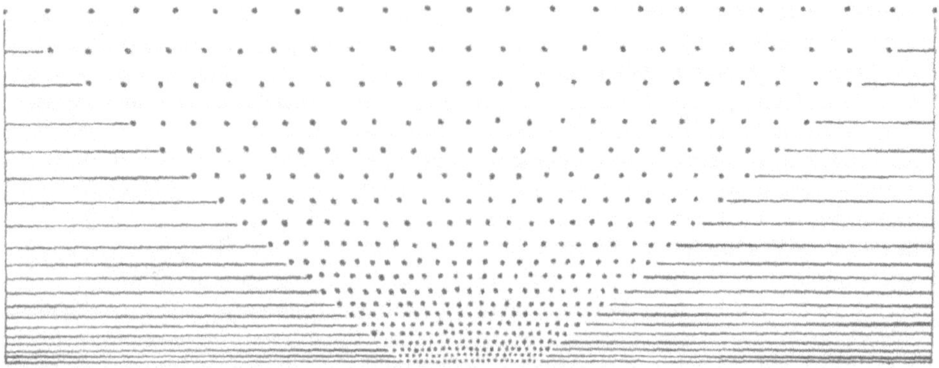

**Figure 11. A parallel-row design (Bleasdale 1967).**

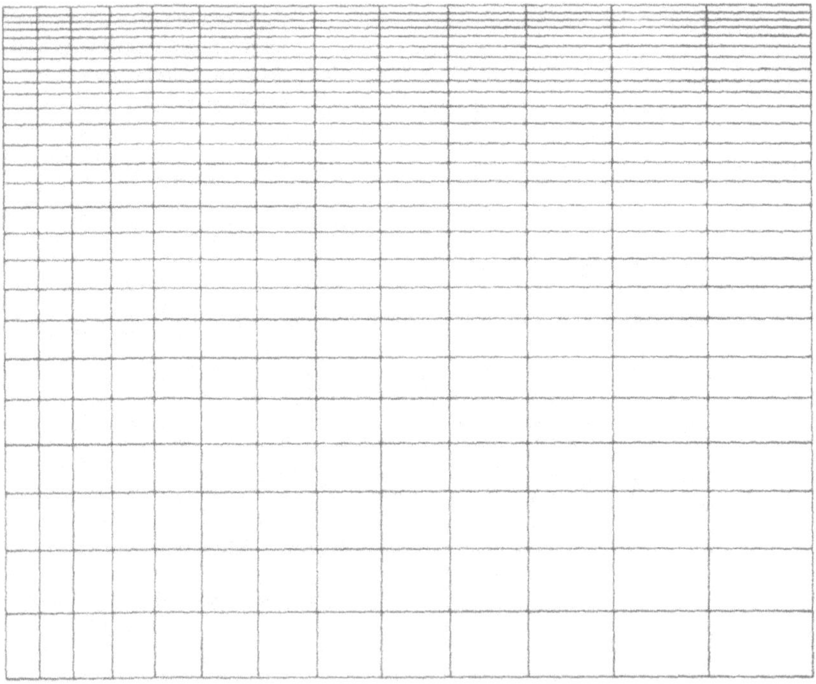

**Figure 12. Two superimposed parallel-row designs. Vertical lines indicate rows of trees; horizontal lines indicate rows of crops**

**Two superimposed parallel-row designs**: In order to examine the effect of independent variation of the densities of two crops parallel-row designs is adopted. It consists of two parallel row designs arranged at right angles with different scales of both the designs to allow for differences in size and desired density between the two crops (Figrue 12).

## General design

As stated above, agroforestry can use same designs as used in single –species agronomic or forestry experiments. However, one or two more unusual designs which are particularly useful for agroforestry are:

i. A nearest-neighbour design

ii. A beehive design; and

iii. 'Y' design

## A nearest-neighbour design

An example of a nearest-neighbour design; it can be seen that each treatment occurs next to each other treatment (i.e. to the right, left, above or below) on four occasions. Thus any deleterious or beneficial effects of one treatment on its neighbours should balance out. This may permit some reduction in the area needed for guard rows, although it is still preferable to have an adequate guard row in order to minimize such interactions. More complex designs taking account of the relative positions of the treatments are to be found in Freeman (1979).

| 1 | 2 | 3 | 4 | 5 |
|---|---|---|---|---|
| 4 | 5 | 1 | 2 | 2 |
| 2 | 3 | 4 | 5 | 1 |
| 5 | 1 | 2 | 3 | 4 |
| 3 | 4 | 5 | 1 | 2 |

**Figure 13. A nearest-neighbour design**

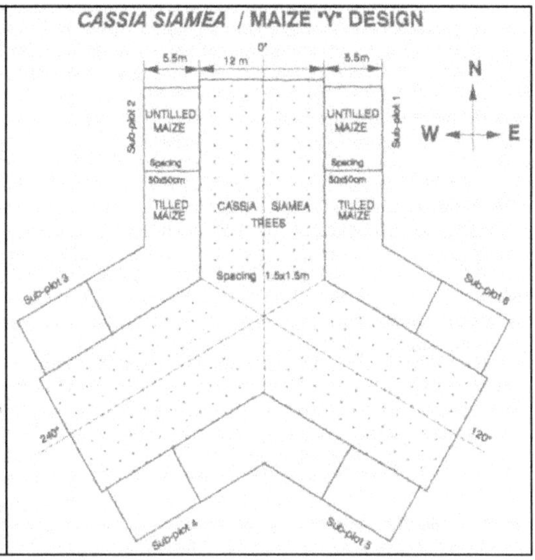


**Figure 14. Shape and dimensions of the 'Y' design plot**

## Y design

The kind of field layout used for tree-crop interface (TCI) investigations i.e. component interaction studies will depend on the precise research objectives and the resources available, including the extent and nature of the site. Orientation may have a significant role at TCI because of the shading effect of tall trees on the understorey crop component. Therefore, orientation as a factor may also be considered (figure 14).

## Beehive Design

Beehive designs suggested by Martin (1973) are used for competition experiments. In these designs there are two kinds of species of plants that are intermingled on a triangular format in such a way that there is a wide range of competition (Figure 15).

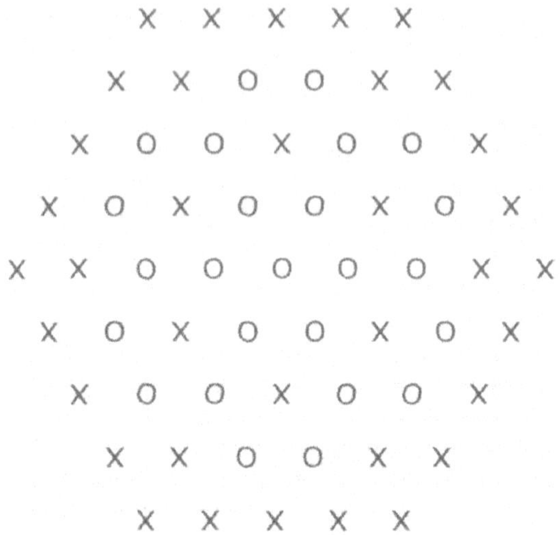


**Figure 15. A beehive design**

**Conclusion:** Designing is one of the most important components for the success of agroforestry. Type of design to be selected depends on various factors such as tree species, crop, topography, slope, cropping systems and type of interactions to be studied. Overall designing and analysis of agroforestry experiments provides information on effect of tree component on soil, climate and crop growth of the agro ecosystems.

# Chapter - 5

# Tree Management in Agroforestry

***Manish Kumar*, Abhishek Kumar and Manish Kumar***

Traditional agroforestry system is practiced in India since time immemorial. In agroforestry, trees are managed for production of timber, fuel wood, fodder, fibres, fruits and other household needs for which trees are lopped or pollarded by the farmers according to their needs. A better understanding of the tree management techniques may leads to greater adoption of agroforestry in small land holding. Agroforestry systems integrate crop production and tree production on to the same area. The trees are planted and maintained in spatial arrangement and temporal sequence than in conventional plantations to minimize the resources competition to support economical crop production. Tree architecture plays an important role in deciding the growth and performance of under storey crops. Crown spread not only affects the soil properties and micro environment but also performance and yield potential of associated agricultural crops. The agroforestry components interact with each other to influence the microclimate and soil conditions (Harper, 1977). The microclimate under agroforestry is somewhat modified due to presence of trees, and hence the growth response of understory crop may be different from sole cropping system (Kumar et al. 2018; Sharma et al. 2002). Manish et al., (2018) observed the decreasing trend of light interception under the tree canopy with significant difference of about 5-8ºC of soil temperature under and outside tree canopy of *Prosopis cineraria* agroforestry in arid region. The quality of solar radiation transmitted by tree canopy decides growth and productivity potential of the field crops (Fischer and Palmer, 1984). Major agricultural crops grown under trees are shade sensitive and are adversely affected due to shading. Shade intensity and duration both will have strong negative effects on the performance of understory crops (Singh, 1994). Therefore, the biophysical science of tree management of the woody components in spatial arrangement and temporal sequence with annual crops is essential to enhance the potentials of system productivity in terms of the biomass and crop yield. Tree management strategy is a set of prescribed methods of manipulating the growth of trees so as to satisfy the objectives of the grower. The main objective of tree management in an agroforestry system is to maximize the value of tree production with minimizing the agricultural production significantly.

The trees should be managed in such a way that the agricultural crops do not suffer from inadequate levels of light, soil moisture and plant nutrient due to competition from the tree components. Pruning, Lopping, Coppicing, Pollarding and Thinning are some of the vital management techniques which are applied to trees and shrubs in agroforestry systems. While, root competition can also be reduced by certain management techniques applied to tree roots such as root pruning etc.

## Tree-Canopy Management in Agroforestry

In agroforestry, tree canopy management is essential operation to reduce the negative interaction among the component and enhance the productivity of agricultural crops. Along with the productivity enhancement, the quality of produce also improved by adopting the canopy management. In agroforestry, manipulation of the canopy of multipurpose tree species is essential operation for integration on the farmland with crops to make it compatible and suitable with associated crops. This is because of following reasons.

1. Canopy size regulates the PAR (Photosynthesis active radiation) in agroforestry, and decides the how much light intensity, interception and transmission of light to the understory crops.
2. Scientific canopy management practices viz., coppicing, pruning, pollarding, lopping and partial crown removal not only affects foliage and branch but also modifies physiological parameters and resource utilization pattern.
3. Canopy management practices helps to maintain hygiene condition by removing dead and diseased branches and also to prevent the spread of diseases in plantation.
4. Canopy management modifies microclimate environment through light, temperature and moisture stress inside and outside the trees.
5. Canopy management plays an important role in managing shade under agroforestry.

Light regulation is major limiting factors which regulates the crop productivity of intercrop in agroforestry system. The most of agricultural crops grown under agroforestry trees are shade sensitive and adversely affected by shading due to the presence of tree canopies and consequently, crop yield reduction. The quantity as well as quality of Photosynthetic active radiation (PAR) intercepted and transmitted by tree canopy decides the growth and productivity potential of the agricultural crops. Temperature and moisture stress is also modifies by the canopy management through circulation of air and water. Rainfall partitioning by tree canopies into throughfall (TF), stemflow (SF) and interception loss (I), which have very significant effect on water balance and the nutrient cycle (Kaushal et al., 2017).

## Basic Principles of Canopy Management

- ❑ To make best use of the land.
- ❑ To make maximum use of available solar radiations for increased productivity.
- ❑ To make convenience in carrying out the cultural practices.

## Purpose and Prerequisite of canopy management

**Purpose** - To optimize the plant model in such a way that canopy is suitable and compatible with the associated arable crops in any agroforestry system by manipulation of resource utilization pattern. The pre-requisite conditions of canopy management in agroforestry are:

1. Full understanding of the crown architecture and growth.
2. Nature of trees.
3. Nature of the companion crops.
4. Objective of agroforestry system.

## Canopy Management Techniques

The most important management techniques with regard to agroforestry trees are:

1. **Pruning**- Pruning of tree involves the removing specific branches or stems to benefit the whole tree. It involves removing of dead, damaged and diseased branches to prevent insect & decay organisms from entering the tree.
2. **Coppicing**- Coppicing involves cutting of the tree at 15 cm from the ground level to obtain flush of new shoots.
3. **Pollarding** - Pollarding is a method of pruning in which trees were cut at 1.0 m height from the ground to obtain flush of new shoots. It keeps trees and shrubs smaller than they would naturally grow. It is normally started once a tree or shrub reaches a certain height, and annual pollarding will restrict the plant to that height.
4. **Lopping** – Tree lopping is the process of trimming various sections of a tree. Lopping may involve clearing away branches with different intensity or even shortening trunks.
5. **Thinning**- A felling made in an immature stand for the purpose of improving the growth and form of trees that remain, without permanently breaking the canopy.
6. **Partial crown removal** - It refers to the direct alteration of tree crown. It ensures proper light penetration to the understory crops. Light penetration is an important factor to determine the extent of crown removal.
7. **Root pruning**- It involves the removal of roots after certain distance to avoid competition with other plant.

# Pruning

Pruning is a selective removal of parts of a plant such as branches, buds, or roots. It is a powerful approach to regulate above ground competition in agroforestry system. Pruning is the most common tree management practice in agroforestry that improves the light transmitting to intercrop and increase the crop production. It is a silvicultural practice to build wood creation, which enhance tree form, shape, structural integrity and possibly uses to get poles and firewood without decrease in wood productivity. It includes removal of live or dead branches or multiple shoots from the tree. Pruning of tree component is a great way to deal with to regulate light, nutrients and other resource competition (Frank and Eduaro, 2003). The pruning treatment can effectively change the microclimate under canopy (such as increased air temperature, soil water content and decrease the relative humidity). Pruning decreased the tree taper and increases the volume along with wood quality. It is mainly of two types: Dry pruning i.e., removal of dead branches and Green pruning i.e., removal of live branches. Another kind of pruning is known as Bud pruning, that involves removal of lateral buds by rubbing off to prevent the development of branches as a measure to obtain knot-free clear bole, it is commonly done in Salix alba and Poplar.

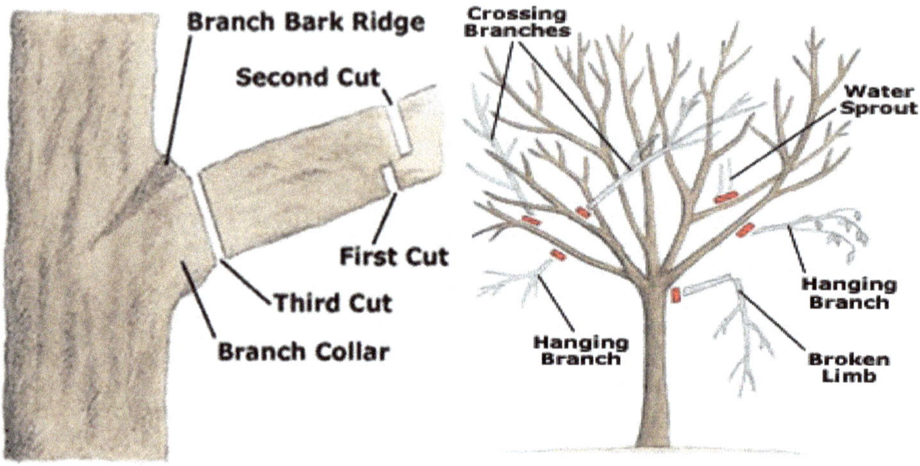


**Figure 13. Pruning techniques in trees**

## The objectives of pruning in agroforestry are:

- Increase the crop production under agroforestry by modifying microclimate.
- Reduction of shade for intercrops near the tree
- To thin the crown to permit new growth and better air circulation.
- To remove weak, diseased, dead, hanging and crossed branches.
- To improve the form and quality of the wood (no nodular wood).
- Early harvest of branchwood for fuel or other use.
- Pruned material used for fodder for livestock and as a mulch production.
- Reduces competition for resources utilization among the components.

The height up to which pruning should be carried out depends upon species, age and local condition. It should be done at least up to height of the adjacent crops if trees are growing in fields. Such pruning facilitates farming operations and reduces competition. In young plantations, it should be carried out up to one half to three-fifth of the total height of the tree. Pruning must be done with an understanding of tree biology. Improper pruning can create lasting damage or even shorten the tree's life. Too much removal of braches may reduce the growth of certain species. During the removal of braches in young tress it should be ensure that there should be at least 4-5 green braches remain uncut, whereas as old trees can tolerate more severe pruning.

## Making proper pruning cuts

Pruning cuts should be made just outside the branch collar. The branch collar contains trunk or parent branch tissue and should not be damaged or removed. If the trunk collar has grown out on a dead limb to be removed, make the cut just beyond the collar. This technique reduces the possibility of tearing the bark and damaging the trunk. Improper cuts may lead to permanent internal decay.

***First cut-*** About a foot from the branch collar, make a cut from the underside approximately a third of the way through.

***Second cut-*** About an inch further out on the branch, cut through the top until the branch rips off. The branch should split cleanly between the two cuts.

***Third cut-*** Make the final cut by placing the saw beside the branch bark ridge and cutting downward just outside the branch collar.

## When to prune?

Pruning of trees begins as early as possible (at the age of two). However, it depends upon the type of plant and the desired outcome. The basic considerations for time of Pruning:

1. **Dead, damaged, diseased, and dying wood:** throughout year, and should be removed as it becomes evident.
2. **Spring-flowering trees:** The trees should be pruned after they bloom for the season since the flower is growing on wood that was produced during the previous years growing season.
3. **Summer-flowering trees:** The trees should be pruned in late winter (after threat of frost) and/or before new spring growth occurs. Their flowers are borne of wood produced that same year.
4. **Trees without flowers:** Should be pruned when they are dormant, typically in mid-late winter or early spring, before new growth had developed. This group includes most deciduous trees, needled evergreens, and many of the broadleaf evergreens.

The best time for pruning is towards the end of the dry season when the work will not interfere with growing crops and when the workload in other agriculture tasks is not so heavy. Four factors determine when agroforestry trees should be pruned: trunk diameter, branch diameter, tree height, and season of the year.

1. **Trunk Diameter-** The object of pruning is to confine the knots created by the pruned branches to a small diameter (four inches) of core wood

thereby producing high quality, knot-free wood on the outer portion of the tree trunk. Consequently, prune trees periodically, never more than once per year, before a portion of the tree trunk grows larger than four inches in diameter.

2. **Branch diameter-** Try to remove branches before they exceed two inches in diameter. Pruning wounds larger than this often take longer than one growing season to close. This results in increased consumption of the tree's energy to close the wound, a longer exposure time for pests to invade the wound, and greater potential to reduce wood quality in subsequent wood growth.
3. **Tree height-** The relative portion of the tree height with branches, or live crown, affects the decision to prune in two ways. The first relates to the health and vigor of the tree and the second relates to the structural integrity of the tree and quality of its wood. Branches support the largest proportion of the energy producing part of the tree, the needles or leaves. Removing too many branches removes leaves, weakening the tree, reducing growth, and making it more susceptible to insect and disease attack. During a single pruning operation, remove no more than one-third to one-half of the total crown. Branches on a tree trunk influence the form and development of the trunk. Trees with less than one third live crown have poorer wood quality and develop weaker stems. Therefore, maintain a live crown of no less than one-third of the tree height.
4. **Season for pruning-** The best time to prune living branches is late in the dormant season or very early spring before active growth begins. Correct timing gives the tree a full growing season to close the pruning wound.

**Height of pruning**: The height of pruning in agroforestry will depends on following factors:

a. Nature of tree i.e photosynthetic efficiency of the tree.
b. Nature of the companion crop.
c. How long the acceptable log length is in the area.
d. Availability of the specialized equipment which allow cheap and safe high pruning.

## How Much Should Be Pruned?

The amount of live tissue that should be removed depends on the tree's size, species, and age, as well as the pruning objectives. Younger trees tolerate the removal of a higher percentage of living tissue compared to mature trees. Generally, no more than 25% of the crown should be removed at once, and less for mature trees. The removal of even a single, large-diameter limb can result in significant canopy loss and can create a wound that the tree may not be able to close. Therefore, care should be taken to achieve pruning objectives while minimizing live branch loss and wound size. The optimal pruning intensity should involve a trade-off between maximising gains both in crop production and wood quality, and minimizing impacts on tree growth.

## Effect of different pruning intensities on growth and Biomass

Khybri et al. (1992) investigated tree biomass and tree crop interactions under *Grewia optiva* and *Morus alba* and recorded that wheat was mostly affected by *Grewia optiva.* The average reduction was ranged from 28 to 34% depending upon the species. From annual removal of lops and tops from trees, *Grewia optiva* could produce 1.08 t $ha^{-1}$ $yr^{-1}$ of branches and 0.26 t $ha^{-1}yr^{-1}$ of leaves (air dry), while *Morus alba* produces 1.28 t $ha^{-1}yr^{-1}$ of branches and 0.28 t $ha^{-1}yr^{-1}$ of leaves. *Erythrina lanceolata* is a potential agroforestry tree, and its slow growth makes it a good alternative as a support tree for climbing, shade tolerant crops like black pepper or vanilla. Partial pruning (50% of the leaf pruned every three months) seems to be a suitable management practice for the species, leading to higher biomass production and N yield in comparison to total pruning every six months resulted in the lowest biomass production (Frank and Eduardo, 2003). Mehta et al. (2012) revealed that pruning of guava plant done thrice a year *viz.* March, May, October to 50% of shoot length can be recommended for maximization of income from ultra-high density guava orchard with total yield/ha as 34.88 t/ha and 37.24 t/ha during 2009-10 and 2010-11, respectively. Palsaniya et al. (2012) carried out an experiment consisting of five treatments pasture + tree without pruning, 25, 50, and 75% canopy pruning and pure pasture in four year old *Albizia procera* based silvi-pastoral system. They observed significantly higher tree height (4.7 m), dbh (9.6 cm), under 75% canopy over other treatments. Similarly, 75% canopy pruning also recorded higher dry leaf fodder (1 t/ha), fuelwood (1.9 t/ha) and total dry biomass (2.9t /ha) from *Albizia procera* over both 25 and 50% canopy pruning treatment (Palsaniya et al., 2012). Meena (2008) studied the agrisilviculture system among different pruning intensities (no pruning, 25%, 50% and 75% pruning) of *Dalbergia sissoo,* and significantly more tree height was recorded in 25% pruning (7.92 m) followed by no pruning (7.47 m) and 50% pruning (7.33 m) where as 75% pruning recorded the lowest tree height. Handa et al. (2007) conducted field experiment under rainfed conditions of Jhansi, Uttar Pradesh to determine the pruning intensity (10, 25, 50 and 75%) and its effect on growth performances of tree species (*Hardwickia binata, Anogeissus pendula* and *A. latifolia*) and yield of Black gram. The maximum Black gram (141.80 kg/ha) and straw yield (233.20 kg/ha) were recorded with *Hardwickia binata* at 75% pruning intensity. No adverse effect was observed with 75% pruning on tree growth while the intercrop yield was recorded maximum. Noda et al. (2007) conducted a field experiment on *Morus alba* with two pruning height (50 cm and 100 cm) and two cutting frequencies (45 and 90 days) and observed that the cutting height showed a significant effect on total biomass in dry season and leaf biomass in both period. The highest plant production was obtained with lowest pruning height. It was concluded that acceptable dry matter yields were achieved with 90 days and 50 cm of pruning height. Pradhan et al. (2003) reported that maximum rhizome yield of turmeric was, when grown within rows of Leucaena planted at 4×1 m apart and pruned five times/year.

## Lopping

Lopping is the harvesting or cutting of the branch of trees in a more haphazard way to collect the fodder. The branches can be cut from any part of the tree but not too close from the trunk. However, it does not take too many branches at a time. The lopping can be done at any time when fodder is required. Lopping is distinguished from pruning in a way that in lopping the branches are not cut from the base. Also lopping is not always done starting from the lower part of the tree and can be highly random. If any selection of branches is to be made, the principle criterion is often a green and leafy biomass, since lopping is usually done to obtain fodder from branches. Lopping is the most common harvesting technique for tree fodder in many areas. Incidentally, the lopped trees produce new shoots which are annually or periodically lopped for various purposes. *Acacia, Anogeissus, Hardwickia binata, Morus spp., Terminalia tomentosa, Kydia calycina, Grewia optiva, Bauhinia variegata etc.* are annually lopped for leaf fodder. The *Grewia optiva* tree can be pollarded and coppices well, whippy branches are lopped yearly for fodder and fibre. One of the main advantages with this technique is that it allows harvest without actually killing the tree. The optimum lopping frequency to ensure survival and rapid recovery will depend on the species, environment and amount of foliage left. Care should be taken in case of certain species whose growth rate may get adversely affected due to heavy lopping (Figure 16).

**Figure 16. View of lopped and unlopped tree**

Singh et al. (2008) conducted tree management practices of *G. optiva* - lopping and pollarding along with Napier as an under storey grass species were compared with unmanaged plantation and recorded 66% survival of grasses under pollarded tree than 58% in control after 10 year. Grass biomass production increased rapidly from the 2nd year and reached the highest level (25100 kg ha$^{-1}$) in the 5th year in the pollarded tree plots, while production decline was quite rapid in the lopped and grass only plots. Therefore, a combination of pollarded *G. optiva* trees with hybrid Napier can be a successful silvi-pastoral practice for degraded lands and can provide sufficient fodder and firewood on an annual basis).

## Pollarding

Pollarding is a process in which crown of the trees are harvested at least two meter above the ground level, *i.e.*, beyond the reach of cattle in order to get flush of new shoots. It is commonly practiced during the period of extreme forage shortage to produce emergency fodder without killing the tree or greatly reducing its height. In some fruit species, pollarding and pruning of branches is necessary to get new flush to bear abundant flowering and fruiting as flowering occurs only in new sprouted shoots. eg. Ber (*Ziziphus mauritiana*) (Figure 17).

**Purposes:** Pollarding can have several purposes of which some are as follows:

- ❑ Applied in the alley and other agroforestry system to harvest wood, fodder, mulch and other biomass.
- ❑ Pollarding is preferred over coppicing practice in wood-pastures and grazing areas where growth from the ground upwards was less practicable due to the required area for grazing and browsing.
- ❑ To reduce shade for the adjacent crops.
- ❑ Pollarding above head height protects valuable timber or poles from being damaged by browsing animals.
- ❑ Regeneration of the tree crown to promote growth of the trunk for timber or poles.

**Figure 17. A tree before and after pollarding**

### Pollarding management (When and how to done)

Pollarding operation will depend on nature of species and season. Pollarding may be started when the tree is young i.e. 2 to 3 years. The older trees when pollarded may result in the death of the tree, especially if there are no branches below the cut. The season and timing of pollarding of tree is also important. The best time for pollarding of many trees and shrubs is in late winter or early spring. While, least favourable time for pollarding is the autumn, as decay fungi may enter the pruning cuts. Once tree is allowed to resprout after the initial cutting, pollarding requires regular maintenance by pruning when resprout begins.

## How to pollard?

- On a shrub, this might be a one stem cut to a meter high, a mass of stem will grow from the top.
- With a tree, it is more typical to leave a trunk supporting three to five branches and this branches are cut to a desired length i.e. 1 to 2 inches from the main stem.

## Maintaining a pollard:

- Once a tree and shrub is pollarded, continue the annual cycle of cutting.
- Branch should be pruned just above the previous pollard cuts.

A tree that has been pollarded is known as pollard. A tree which has been allowed to grow without being cut as a pollard is called a maiden tree. Pollard trees may achieve much greater life-span than maiden trees because they are maintained in a partially juvenile state. They tend to grow more slowly than maiden trees, with narrower growth rings in the years immediately after cutting. Pollarding is best carried out during the dormant winter months, when the leaves of deciduous trees have fallen, it is much easier to see the shape you are creating and there is less stress to the tree as there is minimal loss of sap. Insect and fungal infection is reduced during cold months, as these possible sources of infection are not active.

The choice of pollarding height and frequency depends upon the desired products. If the main aim of pollarding is production of timber or poles, the top of the tree should be cut as high up as possible, and the pollarding interval should be such that the crown is kept as green and vigorous as possible for the maximum production of trunk wood. An interval of 2-5 years is appropriate in such cases. On the other hand, if the main aim of pollarding is production of fuelwood or fodder, it is better to pollard lower down the tree to facilitate access. Pollarding can then be done more frequently, for instance once in a year. It is advantageous to form a wide "stool" (the part of the tree remaining at the base when it has been cut) in order to achieve a substantial production of biomass. Sometimes, the main aim is to produce staking material, poles for the purpose of construction. In such situations, a wide stool will allow many stems to grow. Initially, a stand may sprout too dense after pollarding, and thinning is then recommended, leaving a suitable number of branches in relation to the size of the stems eventually desired.

Tree species with vigorous epicormic growth may be made into pollards. In these species, removal of the main apical stems releases the growth of many dormant buds under the bark on the lower part of the tree. Trees without this growth will, of course, die without their leaves and branches. Some commonly pollarded species are: *Casuarina spp., Acacia spp., Grevillea robusta, Morus alba, Terminalia arjuna, Leucaena leucocephala, Gliricidia sepium, Ailanthus altissima, Salix spp. etc.* In case of Mulberry, farmers generally start rearing silkworm after 3rd year. After completion of rearing, trees are pollarded to get flush of new leaves for rearing of next silk crops. In Chotanagpur district of Bihar, farmers use pollarding as a management technique for *Terminalia arjuna,* tasar silk worm host plant under agroforestry system. At Ranchi, it was observed that the pollarding of *L. leucocephala* around crop fields in two rows provide highest annual forage yield (Sen, 1984). Live fences of *G. sepium* are pollarded for the production of fodder, green manure and fuelwood (Wiersum and Dirjosoemarto, 1987). In Kenya, intense pollarding of *G. robusta* may be carried out 15-20 times over a period of 50 years. The trunk will continue to grow and widen, and also increase in height of stem. Whenever farmer decides that it is large enough, the trunk is felled and is sold for timber (Poulsen, 1983).

## Coppicing

Coppicing is the process of cutting back of the tree at height of 10 - 50 cm *i.e.* close to ground level to stimulate new shoots from the stump. It is a method of regeneration in certain trees and shrubs which have the ability to resprout after the whole tree has been cut off. Coppicing can be regarded as a method of tree propagation since it can substitute for the task of planting a new tree after a mature one is felled. Shoots sprout from residual stump technically referred to as the stool coppice.

### Advantage

Coppice growth is generally much more vigourous than the normal seedling growth. It is an art of getting more stems per trunk. It usually induces fast growth of vigorous multiple shoots. It is a useful practice when it is desirable to produce large quantities of foliage close to ground level, as in the case of many fodder species. It also helps to harvest fuel wood, poles and small sized timber within a short period. It is the method of less investment and quick returns.

### Purpose of Coppicing

Systematic coppicing is applied as the management technique in the alley and other agroforestry system to harvest wood, fodder, mulch and other biomass and to reduce shade for the adjacent crops. It may be considered an option for tree propagation on soil-conservation structures. In such a situation, coppicing may be done annually, but in other situations, for example regeneration of Eucalyptus for pole production, it may be much less frequent. In that case, an interval of 6-8 years may be more suitable.

a. When it is done to produce mulch, fodder, or small wood, the sprouts can be left to grow and be pruned when needed.

b. When the intention is to produce a new tree, all sprouts are removed except one that is going straight up. This will form new tree.

Not all tree species will coppice after being cut. Some commonly coppiced species are: *Calliandra calothyrsus, Cassia siamea, Cassia spectabilis, Eucalyptus spp., Leucaena leucocephala etc.*

Points to be keep in mind during coppicing:

- ❑ Coppicing frequency depends upon species, environment, and amount of foliage left.
- ❑ Depending upon the tree species used and the crop with which it is grown, coppicing can be done 4 to 5 times in a year (sometimes even more).

- ❑ Trees should not be coppiced in dry period but best time to be done in the dormant season.
- ❑ Trees should be coppiced only when attains good health.
- ❑ Trees should be coppiced when it attains 3-5 m height.
- ❑ Cutting height should be 30-50 cm above the ground level.
- ❑ Cut should be angled.
- ❑ Bark should not be damaged while coppicing.
- ❑ Young coppice should be kept weed free.

## Thinning

At the time of establishment, trees are usually planted closer in order to obtain straight growth and fewer branches which are desired in any agroforestry system. Later these trees are thinned to the desired spacing. Thinning is a practice that reduces the competition for light, water, and nutrients and thereby improves the overall health and growth potential of a stand. Thinning ensures that evenly spaced quality trees have the best growing conditions. The timing of thinning can vary depending on species and their growth rates. In agroforestry, mechanical thinning is followed which is usually applied in young plantations before the crown differentiation has taken place. It consists of the removal of alternate diagonal lines or rows of trees which reduces the stocking to the desired level.

**Table.1. Some trees species and suitable management practices in Agroforestry systems**

| Tree | Suitable operation |
|---|---|
| *Acacia auriculiformis* | Removal of lower branches and of young plants as a means of improving stem form and of reducing the incidence of multiple stems. It responds well to pollarding. Young trees respond to coppicing better than old trees, but the tree does not sprout vigorously or prolifically. |
| *Acacia holosericea* | Lopping and pollarding are recommended. |
| *Acacia sieberiana* | The trees respond well to lopping, pollarding and coppicing. The seedlings need to be protected from fire and browsing. |
| *Acacia Nilotica* | Lopping, pollarding and coppicing are recommended. |
| *Albizia lebbeck* | *A. lebbeck* coppices well, responds to pollarding, pruning and lopping, and will produce root suckers if the roots are exposed. |
| *Albizia saman* | The tree responds to pruning and coppices well. There is rapid regrowth of lopped and pollarded trees. |
| *Alnus nepalensis* | Pollarding is recommended |
| *Anogeissus latifolia* | The tree produces root suckers, coppices and pollards well, but this exhibit great seasonal variability. Coppicing and pollarding should not be done during the rainy season. Thinning of coppice is necessary. |
| *Balanites aegyptiaca* | Coppices and pollards well and can regenerate after lopping and heavy browsing. |
| *Berchemia discolor* | Coppicing, pollarding and lopping are all recommended. |
| *Boswellia serrata* | It produces root suckers, coppices and pollards well. The power to produce new shoots depends upon the climate and edaphic factors. |

| Tree | Suitable operation |
|---|---|
| *Bridelia micrantha* | Pollarding and coppicing are recommended. |
| *Cajanus cajan* | For hedgerow intercropping, the hedges should be cut at height of 0.5-1 m when the grain crop is fully mature. The hedges can be cut 2-3 times a year in areas where the dry season lasts 4-6 months. |
| *Calodendrum capense* | Coppices and pollarding are recommended |
| *Cassia abbreviata* | Pollarding, coppicing, trimming and pruning are recommended |
| *Cassia javanica* | It coppices vigorously and can be pollarded. |
| *Casuarina junghuhiana* | It produces strong root suckers. Pruning in plantations up to a height of 2-2.5 m is often necessary. Trees respond well to coppicing and pollarding. |
| Celtis australis | The tree pollards and coppices well. |
| Dalbergia latifolia | Coppicing and root suckers recommended. |
| *Dalbergia sissoo* | Pruning and coppicing recommended. |
| *Diospyros melanoxylon* | Pruning, coppicing and pollarding are recommended. |
| *Eucalyptus spp.* | Pruning, coppicing are recommended. |
| *Erythrina abyssinica* | Pollarding and coppicing are suitable. |
| *Erythrina lanceolata* | Partial pruning are suitable (50% of the leaf pruned every three months). |
| *Ficus religiosa* | Lopping and pollarding are suitable. |
| *Gliricidia sepium* | Pruning and pollarding are the main management activities. Pruning at 0.3-1.5 m will stimulate leaf production. |
| *Grewia optiva* | The tree pollards and coppices well, whippy branches are lopped yearly for fodder and fibre. |
| *Leucaena leucocephala* | It is a vigorous coppicer and responds well to pollarding and pruning. |
| Malia azedarach | It does not coppice well from large stumps, but excellent coppice is obtained from trees up to a girth of 0.9 m. The coppicing ability and pollarding makes it suitable for pole production. |
| *Mimosa scabrella* | It is not a good hedgerow species because it does not coppice, though it can be pollarded or pruned effectively. |
| *Moringa oleifera* | It coppices and pollards well. |
| *Piliostigma thonningii* | Lopping, pollarding, trimming and coppicing are suitable |
| *Pongamia pinnata* | Trees coppice well and can also be pollarded and pruned. |
| *Populus ciliata* | It is a light demander, pollards vigorously, has poor coppicing power except in the case of young trees, and produces root suckers abundantly. |
| *Terminalia arjuna* | Pruning is required to remove the weaker shoots when forking takes place due to frost damage. It also produces root suckers and is suitable for pollarding. |

## Root Pruning

Trees, shrubs and herbaceous plants can have root systems that are shallow, deep or both. Competition for moisture and nutrients will be less if one of the two plants growing together is deep-rooted and the other is shallow-rooted. If the annual crop and trees both have shallow root systems, competition can be minimized by pruning roots so they do not occupy the same soil area. Root pruning is usually focused on the roots of trees. Ideally, you will want to prune tree roots every year or two, starting shortly after planting. If you have an existing stand of trees, try to prune part of the root system each year for several years rather than all at once. This will prevent serious damage to the tree. Manual root pruning can be done by forcing the blade of a spade into the soil or by digging a trench. Mechanized root pruning is done with a tractor mounted ripper, coulter or chisel plough, ideally with subsurface knifes attached to sever more roots. How close to the trunk to prune the roots depends on where the crop between the trees is growing. Pruning tree roots inside a trees drip line should be done cautiously to minimize damage to the tree.

## Conclusion

Suitable management of tree canopies in agroforestry, permitting adequate transmission of photosynthetically active radiation beneath canopies has been found more critical for improving productivity of associated arable crops. Appropriate management practices not only regulate biomass production ability and maintain vigour for extended period but also help agricultural crops to perform better under agroforestry intervention. Canopy without proper management not only reduces the productivity of agricultural crops, but nevertheless deteriorates the quality of produce as well. So we need to continuously refine strategies and update the technology package based on the past experiences and failure to make agroforestry more viable, beneficial and widely acceptable option. The gains of agroforestry research with regard to canopy management till date are certainly impressive. However, in order to transform present day agroforestry into high-tech scientific adventure in the future, there is need to continuously update technology in order to attain predictable understanding that will have wider applicability.

# Chapter 6

# Tree Improvement in Agroforestry

***Sneha Dobhal***

India is one of the ten most forest-rich countries of the world along with the Brazil, Canada, United States of America, China, Democratic Republic of the Congo, Australia, Indonesia and Sudan. Together, India and these countries account for 67 percent of total forest area of the world. As the country has a net sown area of agroforestry is 46.84%, it is one of the most important potential areas for tree growing along with the agriculture crop (FSI, 2000). In order to attract farmers toward agroforestry, we should have viable solution which can provide attractive financial returns to the farmers (Chauhan, 2000).

Forests contribute a lot to the ecology and economy of the nations and play a major role in improvement of quality of life of the people. Due to the impact of biotic pressure on our forests, many forest areas spread across the country has been depleted and degraded which is a cause of serious concern. Therefore, an urgent need has arisen for improvement of forests in the country and that can do by 'Tree Improvement'.

Tree Improvement is the science of development and care of forest. It is all about management of forests by protecting, conserving and upgrading forest resources which also include farming of trees used as timber. This stream of science assesses, characterizes and utilizes the genetic diversity of forest tree species and carries out tree improvement of major plantation species through breeding, and appropriate silviculture packages. The main objective of this branch of science is to increase the productivity of our forests and plantations to meet the requirements of the forest fringe communities and forest based industries. Besides this, it also helps in restoring the ecology of degraded areas to expand the forest and tree cover and improve the environmental quality. Tree improvement relies basically on understanding variation that naturally occurs in tree populations. It increases the value of a tree species by 1) selecting the most desirable trees from natural stands or plantations, 2) breeding or mating these select trees and 3) testing the resulting progeny. The trees involved in this process are referred to as the breeding population. This three-step process is then continuously repeated to further improve the average value of the breeding population.

## Specific Objectives of Tree Improvement in Agroforestry

The specific objectives of the tree improvement programme in agroforestry are to be reviewed regularly and include the following:

- Determine priority tree species for afforestation, reforestation, agroforestry and other tree growing needs
- Determine genetic variation of the priority species
- Select most desirable trees based on the traits and characteristics to form breeding populations
- Study the breeding biology and major constraints for the priority tree species
- Develop breeding strategies of the priority species
- Assess heritabilities and genetic gains of key traits in the priority species
- Develop seed and propagation technologies including protocols for tissue culture
- Design, establish and maintain seed orchards and seed production areas to optimize germplasm production
- Establish gene banks for conservation of germplasm and forest genetic resources
- Ensure regular performance assessment of trials focusing on relatives to increase the effectiveness of the selection process, and juvenile mature correlations
- Determine periodicity and tree seed yields
- Undertake studies on tree seed science and technology

## Component of General Breeding Strategy

Surveying and identification of superior phenotype

↓

Establishment of clonal bank

↓

Carrying out clonal matings

↓

Carrying out genetic test

↓

Manipulation of time in breeding cycle

↓

Establishment of seed orchard

↓

Production of quality planting stock

↓

Maintenance of broad genetic base

↓

Management of inbreeding in breeding and production population

↓

Continuity in improvement programme

## Tree Breeding Techniques

### (A) Conventional Tree Breeding techniques

1. Plus tree selection
2. Provenance test
3. Hybridization and production
4. Accelerated breeding

### (B) Modern Tree breeding Techniques

**1. Seed Orchards**

**2. Micropropagation**

a. Somatic embryogenesis
b. Organogensis

**3. Genetic Engineering**

a. Agrobacterium mediated gene transfer
b. Biolistic method of gene transfer

**4. Cryopreservation and *in-vitro* storage**

**5. Use of molecular markers**

## (A) Conventional Techniques

### 1. Plus tree selection

Preliminary reporting of the outstanding tree called candidate tree and the fFinal appraisal and approval is known as plus tree. There are basically three method applied for the selection of a plus tree

- Comparison tree selection
- Baseline selection
- Regression method

## Comparison tree method of selection

First selection of candidate tree is screened for traits in relation to few surrounding trees. Comparison tree should be selected from dominant or co-dominant crown with similar age and site conditions, approximately within 100m range from the candidate tree. Superiority of the candidate tree over the average of the comparison tree is worked out for each trait. Candidate tree is designated as plus tree if it proves superior to the comparison tree; otherwise rejected. In candidate tree selection, selection should be done from even aged natural stands. The candidate tree should be 5% superior in height and 20% in dbh.

## Baseline Method

It is the selection made simultaneously for all the characters but rejecting all individuals that fail to meet the minimum selection standard for any one trait. Method based the fixation of minimum acceptable standard level for each trait. Individual are located and their values for traits of interest are compared with the minimum selection standard, below which individual are eliminated regardless of their acceptability in other characteristic. This method is useful in uneven aged stands when the comparison tree adjacent to the candidate trees is not available. The candidate tree in this case is not compared with the surrounding tree.

## Regression selection system

The most useful method of tree grading for the uneven-aged or mixed species type stands. A regression selection system is built by sampling a number of trees for desired characteristics, such as volume growth on a given site, and then plotting them against age. No need for comparison tree in this method. Suitable for all aged or mixed species.

### 2. Provenance testing

Provenances are the original geographical area from which seeds or other propagules was obtained. Seeds are collected from a number of widely scattered stand and the seedling are grown under similar condition to select a seed source for better growth and adaptability.

The ideal sequence of provenance testing is

- Range wide provenance phase

- ❑ Restricted provenance phase
- ❑ Provenance providing phase

***Range wide provenance phase***

This phase is the testing phase for all the geographical variation available within the species. 10-30 provenances are selected and tried during this phase. The trial is laid out on small plot size.

***Restricted provenance phase***

On the basis of this phase, 3-5 provenances are selected that are found suitable for the site under site.

***Provenance providing phase***

To confirm the result of restricted provenance phase, this trial is laid out under normal plantation condition.

**3. Hybridization**

Hybridization as a broad term refers to natural or man-made crosses of individuals of unlike genetic constitution. A tree species valuable for its timber characteristics, but lacking in resistance to a potent disease or insect pest, may be combined with a closely related resistant species to produce a hybrid including all the desirable characteristics. A species desirable for a certain region from a forestry standpoint but lacking in winter or summer hardiness may be combined with a hardy species to produce a hybrid adaptable to the region.

**4. Accelerated breeding approaches**

Essential requirement for generation interval can be shortened in tree breeding programme. Accelerated breeding, reducing the time required to mate parent tree and collect seed from them once they are selected. Involve understanding about the basic physiological processes involved with maturation and in reproduction in tree and how this process can be manipulated to promote flowering.

## Modern Techniques

### 1. Seed Orchard

A seed orchard is a plantation of genetically superior tree isolated to reduced pollination from genetically inferior outside source, and intensively managed to produce frequent abundant, easily harvested seed crops (Zobel *et. al.*1958). Kinds of seed orchard are:

- ❑ Clonal seed orchard
- ❑ Seedling seed orchard

**Clonal seed orchard**

In clonal seed orchard the vegetative material of selected phenotypes or plus tree is planted in area with good isolation, under condition favouring flowering and fertilization, and managed for the production of maximum amount of seeds. Clonal seed orchards are of two type:

- ❑ Clonal seed orchard without progeny testing
- ❑ Progeny tested clonal orchard

### Without progeny tested

This consists of tree selection in the wild planted stand, grafting of scion and establishment of seed orchard. No progeny testing is carried out. It is appropriate method if natural population is in danger of extinction or produces little seed in natural habit and for gene conservation.

### Progeny Tested

Orchard of this kind contains large number of clone obtain through plus tree selection and grafting. Progenies are tested after open pollination and orchard is rouged on the basis of progeny test result in which only the best clone are retained.

### 2. Seedling seed orchard

In the seedling orchard, progenies from open or controlled pollination of selected phenotypes are planted at normal plantation spacing. The identity of families is maintained in order to allow to rouging among families and among the individual within families, bases on their phenotypic performance.

### 3. Micropropagation

This refers to *in-vitro* plant propagation methods. The principal approaches are axillary budding (actually a miniaturization of propagation with cuttings). Induction of adventitious buds on non-meristematic tissue (i.e. inducing a shoot where one would not normally develop). Somatic embryogenesis (where individual cultured cells or small groups of cells undergo development resembling that of the zygotic embryo). Work done with some crop species indicates the possibility of encapsulating somatic embryos to form artificial seeds which can then be handled like conventional seeds.

### 4. Genetic engineering

This comprises the insertion of novel genes into a plant or else the modification of existing genes through manipulation of the DNA molecule. It is important that genetically engineered genotypes be of high quality with respect to other traits as well. The clonal test is the most logical basis for the integration of genetic engineering into traditional tree improvement programmes. Genetic engineering is most appropriately conducted with species for which breeding programmes are advanced and clonal forestry can be realistically contemplated.

### 5. Cryopreservation and *in-vitro* storage

This comprises the maintenance of cells, tissues or organs in cultures where growth is slowed (e.g. by the reduction of light, temperature or nutrients) or suspended (by immersion in liquid nitrogen).

### 6. Use of molecular markers

The use of molecular markers involves the examination, using sophisticated biochemical techniques, of variations in cellular molecules such as DNA and quantification of genetic variation. The major current value of molecular markers lies in long-term strategic research; marker studies are making great contributions to advances in the understanding of basic genetic mechanisms and genome organization at the molecular level.

## Features of Tree Species Suitable for Agroforestry

- ❑ Fast growth – 20-25m$^3$/ha/yr (Singh *et al.*, 2001)
- ❑ Straight clean bole
- ❑ Leaflessness during winter
- ❑ Multiple uses-Pulpwood, packing cases, poles, timber, etc.
- ❑ Compatibility with agricultural crops
- ❑ High economic returns – Rs 88,749/acre @12 percent interest in seven years (Dillon *et al.*, 2001)
- ❑ Short rotation period (6-8 years)
- ❑ High $CO_2$ exchange rate (Nelson, 1984)
- ❑ High water use efficiency (Dickmann and Keathley, 1996)

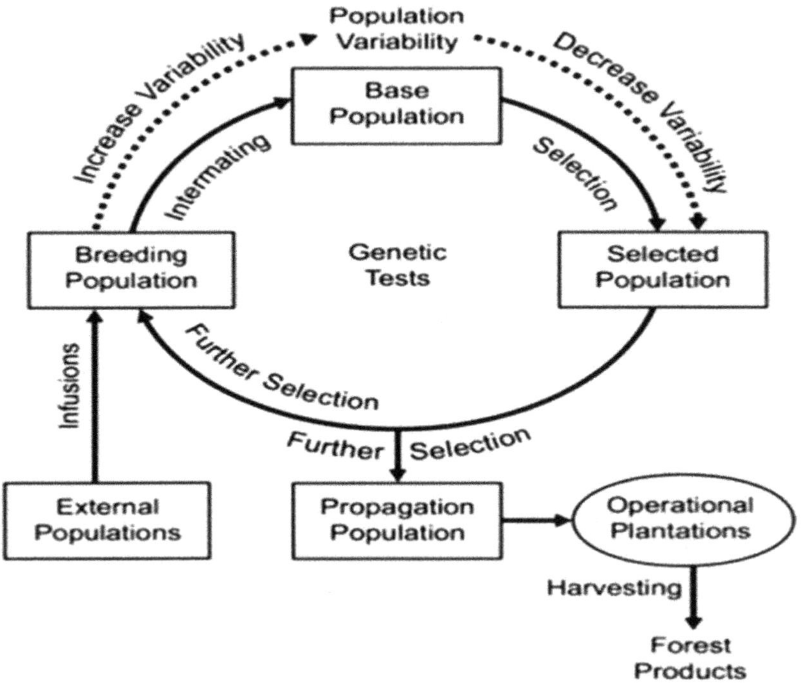


**Figure 18: Cycle of Tree Improvement Programme**

## Climate Change and Tree Improvement

Foresters and tree breeders manage genetic adaptation to climate in commercial forestry by using seed zones and tree breeding regions (Gray et al., 2016). Seed source/provenance research has been described as the most important contributions of forestry to biological sciences (Matyas, 1996). Seed source is defined as the geographical area from where seeds or planting material has obtained to test in new region (Zobel and Talbert, 1984). The natural variation presents in different geographical regions induces variation in the plant characters and productivity in a particular species (Jones *et al.*, 2001), as a consequence, plants respond differently in new region compared to the original geographical region (Lande and Shannon, 1996). Another, strategy that can to guarantee adaptation for climate change is to select best performing tree from stands subjected to climate change impacts previously. Indeed, a particular seed source may exhibit superior plant characteristics and adapt to changing climate in new region compared to other seed sources, because the seed source variations are genetically controlled (Mazanec and Mason, 1993). During late twenties, Ledig and Kitzmiller (1992) suggested genetic strategies for reforestation in the face of global climate change. In general, improvement in plant functions through seed source selection, testing and crossing are of great significance for meeting afforestation needs, which can provide greater climatic and economic benefits such as controlling soil erosion, mitigating climate change, improving carbon stock and the provision of fuelwood, fodder, fruit and timber (Zobel and Talbert, 1984; Oliveira *et al.*, 2015).

Some of the important tree species used in agroforestry with applied improvement method is given below:

### i) Poplars (*Populus* spp.)

Various species and clones of poplars have been introduced from Germany, UK, USA, Netherlands, Japan, China, and Australia. In India, gene bank of poplar clones are being maintained at germplasm bank for future breeding trials as well as a source of clonal material for stool-bed and for reference identification. In order provide broad genetic base, to develop new clones which are superior to growth traits and were screened for growth traits continuously. The *Populus ciliata* & *P. deltoides* and its hybrid clones recommended for higher productivity. The superior clones of *P. deltoides* i.e. G3, G48, L34, $S_7C_{15}$, Uday, Kranti and Bahar are that are suitable for the different agroclimatic conditions of northern Indian states like Punjab, Haryana, Uttar Pradesh and Uttarakhand.

### ii) Willow (*Salix* species)

Clones of willows (*Salix* species) have been procured from different countries for testing in farmers fields as multilocation trials and have been found to be promising in the field trials. *Salix* hybrids (F1s) developed through control-pollination by involving outstanding clones. Genetic characterization of introduced clones and hybrids was analyzed using SSRs and RAPD molecular markers also. Supply of willow germplasms under multilocation trial at various locations (26 institutes/ NGO) in the country.

**Figure 19: Wheat with *Populus deltoides* Turmeric (*Curcuma domestica*) with *Populus deltoides***

## iii) Beul (*Grewia optiva*)

Successful grafting and chip budding of the superior genotypes were undertaken on seedling stock of *Grewia optiva*. The grafting is more successful than chip budding in *Grewia optiva*. The grafted plants were also used to establish seed orchard. Micro propagation in the species revealed that it can be easily multiplied from shoot tip explants and genetic transformation has been also worked out. The assessment shows that hybridization between genetically distant families can generate promising breeding material to obtain improved genetic gain. Control crossing (hybridization) has been done by involving superior selection and progeny is being evaluated. *Grewia optiva* is mainly used in fodder, fuelwood and bast fibre.

## iV) Bamboo (*Dendrocalamus strictus*)

The clumps from different locations were selected and tested at different locations. Cheap, simple and easy technology for its multiplication has been developed. Seed of different families was collected from different locations, revealed significant differences among the different families for the seed and seedling parameters. 85 per cent survival rate was observed in *Dendrocalamus strictus*. The survival rate of sapling was 90 percent after 6 months with an average of 4 tillers per sapling in *Dendrocalamus strictus*. The development of rhizomes was quite encouraging with an average of five rhizomes per saplings.

Tissue culture that to get rapid multiplication of *Dendrocalamus srtictus* L. through in vitro regeneration from juvenile explants using tissue culture technique for large scale multiplication of the plantlets in which farmers can get true to type plants with all the advantages of vegetative propagation (clonal propagation).

## v) Chinese Bamboo (*Phyllostachyus reticulata*)

Rhizome cutting with buds planted just below soil surface gave 100 per cent success. Spring season was found to be the best season for vegetative propagation.

### vi) Chirpine (*Pinus roxburghii*)

Cones above 16 cm length were found best for seed germination and seedling growth in the nursery. Tree height is more responsive to family and within family selection than diameter at breast height (dbh) in the species. Cleft grafting having scion and stock of the same diameter was standardized which gave success in both open and under partial shade. In vitro propagation studies in *Pinus roxburghii* has been done.

### vii) Seabuckthorn (*Hippophae rhamnoides*)

The species can be easily propagated vegetatively from stem cuttings taken mid-February. Summer stratified seed gave more than 70 per cent germination when sown during the first week of June.

### viii) Harar (*Terminalia chebula*)

High yielding trees of Harar (*Terminalia chebula*) have been identified the grafting technique.

### ix) Toona (*Toona ciliata*)

There is vast potentiality for improvement on the basis of individual tree selection. Pollination mechanisms include open pollination facilitated by common thrips also.

### x) Khair (*Acacia catechu*)

For higher katha content main emphasis be given to specific gravity, DBH, number of primary branches and pod width in addition to the locality aspect (Southern and Western aspect) and tree form (leaning and forked tree). Hence the selection for desirable traits in combination suggests better prospects for improvement.

### xi) Robinia (*Robinia pseudoacacia*)

Mutation studies through physical and chemical mutagens revealed that lower doses of continuous and fractionated gamma irradiation and hydrazine hydrate stimulated the germination and survival of seedlings. The treatment of dormant shoot cuttings treated with lower doses of gamma radiation above (1to2 KR) lead to the reduction of thorn size.

### xii) Neoza Pine (*Pinus gerardiana*)

Cleft grafting techniques on two to three years old seedlings with partial shade in the species gave 70-80 per cent success. If all biotic interference is excluded, complete closure of the area was observed to encourage natural regeneration.

### xiii) Marinoo (*Ulmus villosa*)

The progenies are highly diverse and should be used for hybridization improvement programmes.

### xiv) Kachnar (*Bauhinia variegata*) and Mulberry (*Morus species*)

Protocols for *in vitro* multiplication of both *Bauhinia variegate* and *Morus alba* have been used.

Other examples are *Celtis australis, Tectona grandis, Eucalyptus tereticornis, Prosopis juliflora, Leucaena leucocephala, Acacia tortilis* and Maharukh *(Ailanthus excelsa)* have also been used.

## Problem Peculiar to tree breeding and improvement

- **Time:** require many year to produce the desired traits like volume wood maturity, resin etc. they take years to bear flower and seed
- **Necessity of seed:** needs to be concerned with seed production and effort to experiment on the stimulation of flowering and fruiting
- **No come back:** once a plantation is raised by using a certain seed lot, it cannot be replaced in midway by other seed, as it will give high economic loss. However in crop plant another seed can be used in next season.
- Scarcity of basic genetic information about trees

## Future Research Needs/Thrust Areas

The significance of the agroforestry species and narrow genetic base of field plantation, there is a need to undertake breeding and clonal deployment strategy programme at national level. The following are the thrust areas of research that need immediate attention.

- Introduction of new clones from the International Institutes involved in breeding work. There is an ample scope clonal introduction, as germplasm present in India come from only about 25 percent of the native range of this species in USA. There is a need to replace 5-10 percent of the exiting clones every year.
- Exchange of genetic material among different organization institutions taking up tree breeding programme. Research should be undertaken for the identification of site- specific clones by conducting multi-location trials of these clones at different sites in all poplar growing regions.
- National germplasm bank should be established on the pattern of other annual crops.
- Facilities for advanced research on molecular aspects, viz. genetic engineering, molecular markers, DNA fingerprinting (for identification of clones), etc. should be created.
- Research on wood utilization for finding alternative uses of poplar wood is the need of the day. Efforts to improve the quality of agroforestry tree species by some chemical treatments should be initiated.

# Chapter 7

# Agroforestry for Carbon Sequestration and Climate Change Mitigation

*Nidhi Rawat, Raj Kumar*

## Introduction

Agroforestry term is used for land use management that involves mixture of trees with agricultural crops, pastures and livestock to utilize the ecological and economic interactions of the different components (Nair, 1993 and Young, 1997). According to Smith *et al.* (2012) agroforestry is the agricultural practice, which combines crop, trees and livestock and this is one possible alternative for reducing carbon dioxide concentration in atmosphere. Agroforestry is s climate resilient systems which is profitable, sustainable, adoptable, have high CO2 sequestration potential, resilient to extreme weather and helps in reducing the greenhouse gas emission (Figure 20). Agroforestry plays an important role in enhancing the above and below ground carbon sequestration rate and decreases amount of greenhouse gases from agricultural sector (Branca *et al.*, 2013), although agroforestry has multifold ecosystem services like soil nutrient management, control soil erosion, landslide control, biodiversity conservation, and maintenance of air and water quality (Jose, 2009). Agroforestry is important for livelihood improvement of farmers through increasing food production, providing fuel-wood and fodder and maintain soil moisture by providing organic matter through tree litter.

Carbon sequestration in agroforestry systems is important for climate change mitigation which slows down or even reverse the increasing concentration of atmospheric $CO_2$ by storing organic carbon in soil for several years (Klaus Lorenz and Rattan Lal, 2014). The increase in atmospheric $CO_2$ due to fossil fuel emission is contributing to climate change which is major challenges facing the world today.

## Potential Impacts of Climate Change

Ecosystems and their services are severely affecting global climate change, both directly and indirectly. In India the direct impact of climate change would be on plant growth, development and yield, due to change in rainfall pattern and temperature fluctuations. Increase in temperature would have negative effects such as reduce crop duration, increase crop respiration rate, mineralization in soil and decrease fertilization use efficiency. All these could considerably affect crop yield for long run. Following are the major challenges as a consequence of climate change.

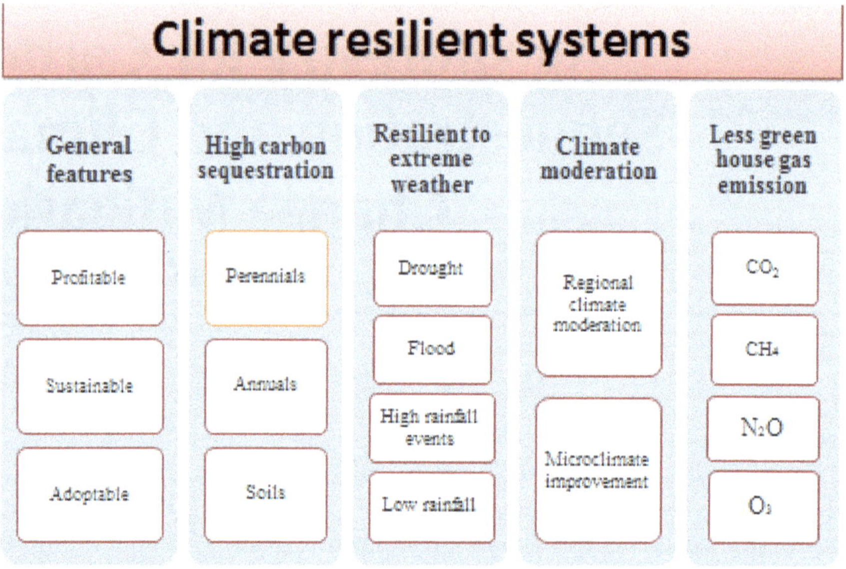


**Figure 20. Conceptual framework showing characteristics of climate resilient systems**

- ❑ An increase in temperature will speed up plants development processes. The shortening of life cycle of plants could have adverse effect on productivity due to earlier senescence which may imbalance the food production between cool and temperate regions.
- ❑ An increase in crop water requirements to meet the increased evapo-transpiration demands.
- ❑ Reduction in the amount of plant available water in most places due to the predicted shortages in water supplies.
- ❑ Increased degradation of land resources from erosion due to the projected increase in extreme events.
- ❑ New pests and vector borne diseases borne that would spread into areas where they were not found previously.
- ❑ Decrease in biodiversity of the world's most fragile environment, such as mangroves and tropical forests.

- Threatening of natural vegetation and fauna.
- An increase in frequency and intensity of extreme weather (drought, floods, etc.) conditions.

## Agroforestry for Carbon Sequestration

A mitigation strategy refers to tools, technologies and practices that widely adapted and will help to reduce carbon emission. Agroforestry system has the potential to reduce the volume of greenhouse gases that rise into the atmosphere. Removing this atmospheric C and storing it in the terrestrial biosphere is one of the greatest options for mitigating the emission of this GHG. According to Intergovernmental Panel on Climate Change (IPCC) report, suggests that land based mitigation – agriculture, forestry, and biomass liquid and solid energy substitutes can be cost-effective land mitigation options. And, it can contribute over the century 94 to 343 Pg C equivalent of greenhouse gas emission abatement, which is 15 to 40 percent of the total abatement required for stabilization. Agroforestry have double potential to address climate change issue by:

1. Greenhouse gas mitigation strategy
2. Sustainable adjustment to changing conditions

Agroforestry system has a particular potential role in the mitigation of atmospheric accumulation of greenhouse gases (IPCC, 2000) All types of land use considered in the IPCC Land Use and Land Use Change and Forestry Report, agroforestry was considered to have the highest potential for carbon sequestration. Emission of $CO_2$ is mainly due to burning of fossil fuels, clearing and burning of forest, expanding and intensification of agriculture. Globally agriculture accounted for 25% of $CO_2$, 50% $CH_4$ and 70 % $N_2O$ emission (Hutchinson, *et al.* 2007). As stated by Makundi *et al* (2004) and several others, estimating the amount of C sequestered by agro-forestry poses unique challenges and have a mitigation potential of 1.1-2.2 PgC in terrestrial ecosystems over the next 50 years (Solomon, *et al.*, 2007). The average carbon sequestration rate of the agro-forestry is 0.2-3.1 t C/ha (Waston, *et al.*, 2000 and Pandey, 2002).

Nevertheless, the IPCC (2000) estimated that 630 million ha of unproductive croplands and grasslands could be converted to agroforestry worldwide, with the potential to sequester 391,000 Mg of C per year by 2010 and 586,000 Mg C per year by 2040. Although the credibility of conceptual models and theoretical benefits has been demonstrated, C sequestration potential is still a little-studied characteristic of agroforestry systems (Nair and Nair 2003). More studies examining how much C can be sequestered or stored in various agro-forestry systems around the world are needed. Several studies and reviews from different regions of the world have discussed agroforestry benefits and limitations for C sequestration (Schroeder 1994; Dixon 1995; Albrecht and Kandji 2003), but only very few deal with comprehensive comparisons of different practices in each eco region. Lal (1999) estimated the potential for sequestering C in the region was, as in most other dry lands, fairly low, between 0.05 – 0.3 Mg C ha-1 yr-1. The estimate, however, included a variety of uncertainties related to future shifts in global climate, land use and land cover, and the poor performance of trees and crops on poor soils in the region. It would also lead to a decline in soil organic carbon (C) stocks and an increase in $CO_2$ emission from soils. Agroforestry systems (AFS) offer solutions to some of these climate change related ecosystem management problems. Agroforestry practices are considered good land management tools to enhance biodiversity (Rois-Díaz, *et al.* 2006; Rigueiro-Rodríguez, *et al.* 2011b) and increases C sequestration, compared with tree-less systems worldwide (Nair, *et al.* 2008, 2009). Intergovernmental Panel on Climate change (IPCC) estimates (IPCC, 2007) that the current greenhouse gases (GHGs) concentrations are 30% more than the pre-industrial level. C is accumulating in the atmosphere at a rate of 3.5Pg (Pg = 1015 g or billion tons) per annum, the largest proportion of which resulting from the burning of fossil fuels and the conversion of tropical forests to agricultural production (Paustian, *et al.*, 2000). Therefore, agroforestry can be one of the tool for climate change adaptation and mitigation throughout the globe.

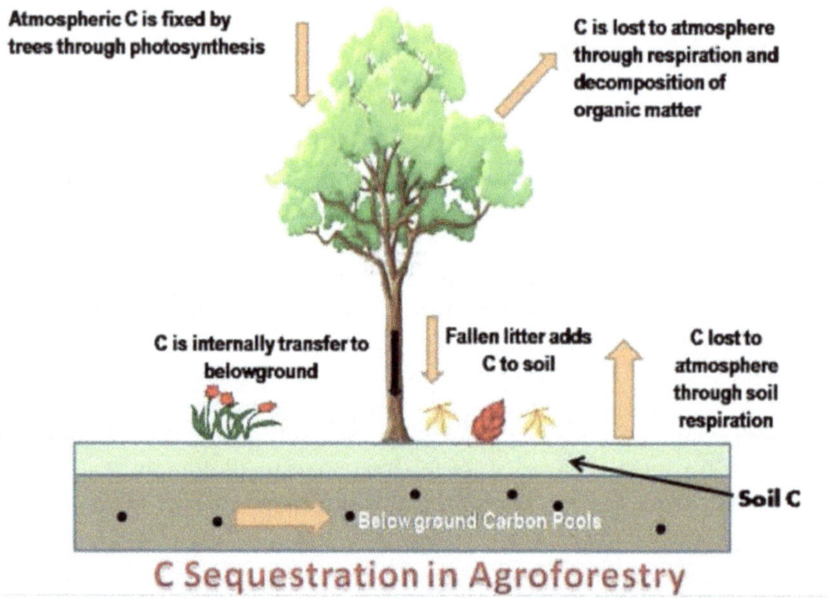


**Figure 21. Carbon sequestration process in agroforestry (Source Newaj et al., 2014)**

## Climate Change Mitigation

Climate change mitigation and adaptation measures at national scale, regional or local level, the policy as well as requirements can vary from country to country. For example, *Faidherbia albida* parklands cover a large area of the Sudano-Sahelian zone of Africa, a region that suffers from soil fertility decline, food insecurity and climate change (Sidaa et al. 2018). The parklands deliver large number of benefits, including fuelwood, soil nutrient enrichment, soil moisture conservation, and higher crop yield underneath the canopy. Its microclimate modification potential may provide an affordable climate adaptation strategy that needs to be explored. *F. albida* trees increase soil N, water use efficiency in wheat, reduce heat stress and increasing yield significantly. As a consequence of climate change heat and moisture stress likely to be more prevalent during 21[st] century, *F. albida*, with its significant impact on microclimate modification, may be a starting point to design more resilient and climate-smart farming systems.

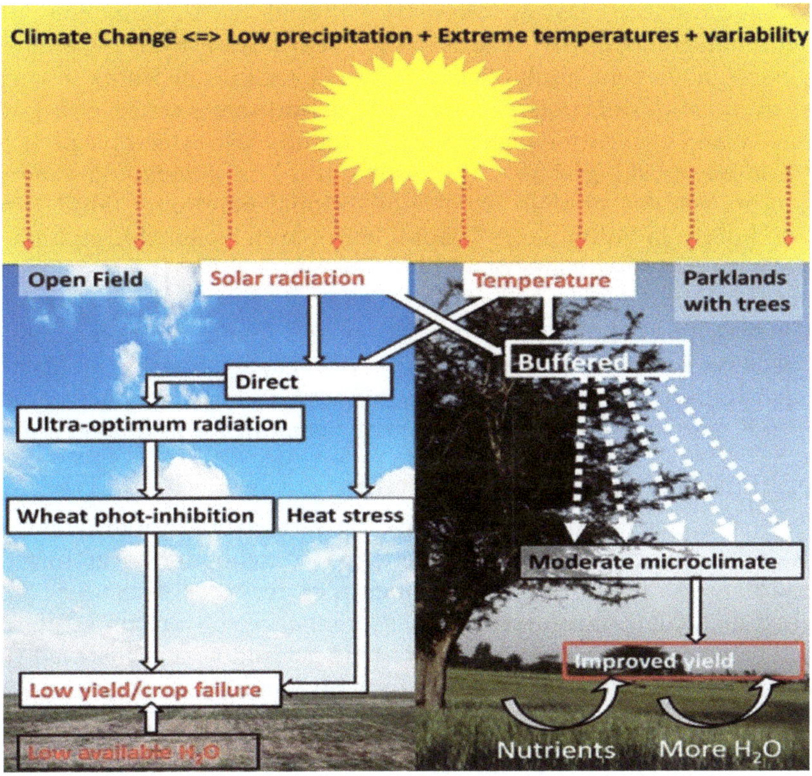


**Figure 22. Climate change mitigation potential of wheat based agroforestry systems (Source: Sidaa et al. 2018)**

## Extreme Weather Events

The trees and agroforestry systems are most suitable farming systems to cope negative impact of extreme weather events vis-a-vis climate change. In this context, Nguyen et al. (2013) in a study in Vietnam have observed that while rice and rain-fed crops suffered over 40 % yield losses in years of extreme drought or flood, while tree-based systems were less affected. The thirteen tree species performed well under the harsh local climate conditions in home and forest gardens to provide income, food, feed and other environmental benefits. The research suggests that maintenance and enhancement of locally evolved agroforestry systems, with high resilience and multiple benefits, can contribute to climate change adaptation. In another study, Simelton et al., (2015) concluded that trees and/or agroforestry have been traditionally used by farmers to cope with various climate-related hazards in Viet Nam. Hence, Integrating trees with crops reduced risks against extreme weather events, and farms with trees had shorter (fast) recovery periods after most natural disasters.

A survey conducted in Central American hillsides after Hurricane showed that farmers using diversification agricultural practices such as cover crops, intercropping, and agroforestry suffered less damage compared to conventional monoculture. The survey, spearheaded by the Campesino a Campesino movement, mobilized 100 farmer-technician teams to carry out paired observations of specific agroecological indicators on 1804 neighboring sustainable and conventional farms. It was found that sustainable plots had 20 to 40 % more topsoil, greater soil moisture and less erosion and experienced lower economic losses than their conventional neighbors (Holt-Giménez 2002). Similarly in Sotonusco, Chiapas, coffee systems exhibiting high levels of vegetation complexity and plant diversity which suffered less damage from Hurricane compared to more simplified coffee systems (Philpott et al. 2009). In Cuba, researchers conducted a farm survey and found that diversified farms exhibited losses of 50 % compared to 90 or 100 % in neighboring monocultures. Likewise, agroecologically managed farms showed a faster productive recovery (80–90 % 40 days after the hurricane) than monoculture farms (Rosset et al. 2011).

## Greenhouse Gas Emission Reduction

Carbon sequestration in agroforestry has received particular attention as a means to slow the increase in atmospheric $CO_2$ concentrations. Agroforestry can increase biomass carbon stocks as the biomass of trees consists of 46–51% carbon (Aalde et al., 2006). Agroforestry can also enhance soil carbon sequestration through enhanced fine root production, rhizo-deposition, and litter fall (Lorenz and Lal, 2014). In a Special Report on Land Use, Land-Use Change and Forestry, the IPCC analyzed various land uses and found agroforestry to offer considerable C sequestration potential, especially in developing countries, with estimated total net annual rates of C stock changes after conversion of unproductive agricultural land to agroforestry was 3.1 t carbon per hectare per year (t C $ha^{-1}$ $y^{-1}$) which is numerically equivalent to 11.4 t $CO_2$ per hectare per year (t $CO_2$ $ha^{-1}$ $y^{-1}$) (Watson et al., 2000; Verchot et al., 2007). While the IPCC report identified some likely potential, it was not supported by a broad and representative review that would have tried to quantify the mitigation potential of agroforestry. There is, therefore, still a lack of comprehensive understanding of the net soil emissions of $CH_4$ and $N_2O$ from agroforestry systems, and whether those emissions are higher or lower than emissions from corresponding agricultural land, even though there are many valuable individual studies that have contributed a wealth of knowledge from individual sites. For instance, Verchot et al. (2006), Baggs et al. (2006) and Dick et al. (2006) found that $N_2O$ emissions from legume-based agroforestry were higher than those from non N-fixing trees or non N-fixing cropping systems. Contrary to that, Rosenstock et al. (2014) argued that, because of the typically large applications of inorganic fertilizers, $N_2O$ emissions from agricultural lands may nonetheless be higher than from legume-based agroforestry.

Kim et al., 2016 in a study observed that Agroforestry (at an average age of 14 years) sequestered 7.2 t C $ha^{-1}$ $y^{-1}$, with biomass C sequestration contributing about

70% and soil C sequestration contributing about 30% of the overall gain. Soils under agroforestry also oxidized 1.6 ±1.0 kg $CH_4$ ha$^{-1}$y$^{-1}$ and emitted 7.7±3.3 kg $N_2O$ ha$^{-1}$y$^{-1}$. Compared with adjacent agricultural fields, soils under agroforestry had lower bulk density, and higher SOC and soil N contents, but pH and net $CH_4$ and $N_2O$ emissions from soils under agroforestry remained nearly the same as under agriculture. Overall, young agroforestry stands contributed to mitigating 27±14 t $CO_2$ eq ha$^{-1}$ y$^{-1}$.

## Micro-climate improvement

Predictions suggest that climatic conditions will become more variable and extreme in much of the arid, semi-arid and sub-humid tropics (Rosenzweig et al., 2004; IPCC, 2007). Simulations suggest that some parts of world will experience increased temperature, a 5–20% increase in rainfall between December and February and decreased rainfall between June and August by 2050 (Hulme et al., 2005). Most of the crops, usually grown as monocultures, are expected to be most affected by increased temperature, while drought is expected to cause widespread famine. Smallholder farmers may combat climate change by reverting to more natural systems which provide beneficial ecological functions (Gregory and Ingram, 2000). This suggestion is particularly pertinent for farmers who lack financial or material resources to adopt technological solutions such as irrigation (Eakin, 2000).

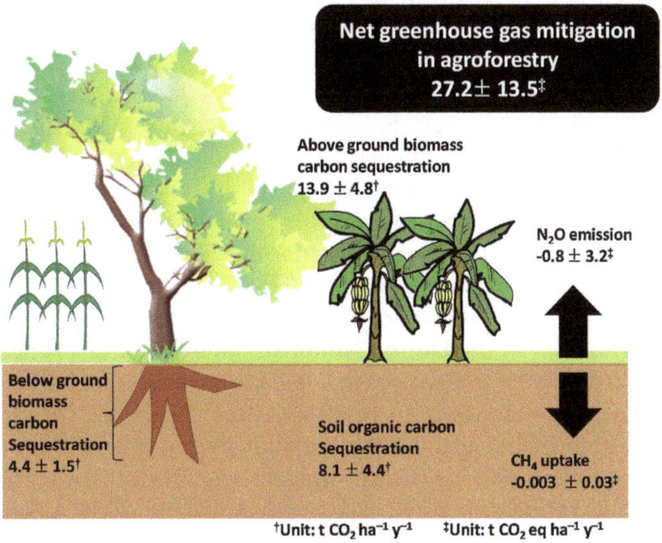


**Figure 23. Changes in the rates of carbon sequestration, methane (CH4) and nitrous oxide (N2O) (Kim et al., 2016)**

Climatic variation has a major impact on crop performance as extremes affect growth, development and yield in a wide range of species (Slingo et al., 2005). Some studies suggest that there are thresholds for temperature and atmospheric saturation deficit, above which physiological and developmental processes and yield become increasingly vulnerable (Challinor et al., 2005; Porter and Semenov, 2005). Air temperature is important as excessively hot conditions during critical developmental stages may greatly reduce yield; thus, temperatures exceeding 30 °C may induce pollen sterility and reduce seed or fruit yield (Porter and Semenov, 2005). Shade from overstorey trees may ameliorate climatic conditions for understory crops, increasing growth and productivity (Gregory and Ingram, 2000). Agroforestry is resource conservation system which may provide a financially viable way of protecting crops in areas where microclimatic factors regularly exceed the optimal range (Lin, 2007). The importance of microclimate in determining crop performance is well documented and has been incorporated into simulation models to predict crop yield over a wide range of environments (Ong et al., 1996, 2006, 2007). Overall tree helps to maintain temperature in below canopy during high and low temperature events. Incorporating tree in agriculture landscape also maintains soil temperature, relative humidity, wind speed, solar radiation, and rainfall and soil evaporation.

## International Policy and Cooperation in Climate Change Mitigation

Under the Kyoto Protocol's Article 3.3, agro-forestry was recognized as an option of mitigating GHGs. Since then, the C sequestration potential of agro-forestry systems has attracted greater attention from both industrialized and developing countries. It is attractive because of its applicability to a large number of people and areas currently in agriculture, as well as its perceived potential for reducing pressure on natural forests. Clean Development Mechanism (CDM), defined in Article 12 of the Protocol adds the attractiveness, because the CDM provides for Annex I Parties (industrialized countries which have emission reduction goals) to implement project activities that reduce emissions in non-Annex I Parties (developing countries), in return for certified emission reductions (CERs) (UNFCCC, 2007). Since agroforestry is traditionally and widely practiced in developing countries, it is feasible/easy options for both developing and developed groups of countries to start as mitigation projects under the CDM. Climate change alters the physical and biological properties of the environment by which living beings are negatively impacted. The international response to climate change started in full with the establishment of the United Nations Framework Convention on Climate Change (UNFCCC) in 1992. Five years later, 159 countries signed a treaty called the Kyoto Protocol, which commits the 38 signatory developed countries to reduce their collective greenhouse gas (GHG) emissions by at least 5% compared to the 1990 level by the period 2008 – 2012. The agreement came into force on February 16, 2005, following its ratification by Russia on November 18, 2004. As of April 2007, a total of 169 countries and other governmental entities have ratified the agreement. A unique characteristic of the Kyoto protocol is that it allows the amount of $CO_2$ sequestered by forests to be counted towards emission targets. Tropical forest conversion contributes as much as 25 % of net annual $CO_2$ emissions globally (Palm et al. 2004). The Kyoto Protocol establishes that land use, land use change and forestry (LULUCF) activities such as afforestation, reforestation, and deforestation (Article 3.3), and forest land management, crop land management, grazing land management, and revegetation (Article 3.4) can be used to meet the greenhouse gas (GHG) emission reduction goal (UN, 1998).

Considering the mitigation strategies to climate change will be effective only if adopted widely by the producing sectors, which requires a lot of communication and a huge degree of cooperation between scientists and farmers. Since the changing climate is not entirely in our hands, we are not exactly sure about what will happen. Thus, it is important that our mitigation and adaptation measures need to be flexible, easy and ready to adjust to be adaptive by giving high attention for natural resources on which majority of people depend. Therefore, policymakers are encouraged to discuss strategies to minimize impacts of climate change. Climate policy discussions have progressed since 2010 and many countries have pledged to limit or reduce their greenhouse gas emissions by 2020 (Hohne, N. *et al.*, 2012).

## Conclusion

Agroforestry systems offer a great opportunity by acting as sinks for atmospheric carbon while helping to attain food security, increase farm income, improve soil health and discourage deforestation. Agroforestry can help to achieve climate change mitigation and adaptation while at the same time providing livelihoods for poor smallholder farmers.

# Chapter 8

# Ecosystem Services of Agroforestry

*Sudipta Ramola, Himanki Dabral*

## Introduction

Globally agroforestry has been in practice for centuries. In very simple words agroforestry can be understood as combination of trees, agricultural crops and/or animals grown together as a multidimensional and multifunctional landscape that offers ecosystem services, environmental benefits and economic commodities (Jose, 2009; Dhanya et al., 2014). It has emerged as a multifunctional, viable land-use option and solution to various global environmental issues and problems such as loss of biodiversity, loss of ecosystem services, deforestation, global warming, land degradation and environmental pollution. Combination of trees and agricultural crops leads to better rural livelihoods, higher biomass production, reduced soil erosion, reduced nutrient leaching, increased biodiversity, enhanced soil fertility, improved water quality, enhanced biodiversity, increased aesthetic value, enhanced carbon sequestration and alleviated resource use pressure on conservation areas (Ong et al., 1996; Palma et al., 2007; Garrett and McGraw 2000; Garrity, 2004; Williams-Guille´n et al., 2008; Nair et al., 2009).

The interaction and exchange of matter and energy between abiotic and biotic components of an ecosystem generates ecosystem services (Singh, 2002). Ecosystem services on one hand are the result of various ecological processes within an ecosystem whereas on the other hand these are ultimately the cause of an efficient ecosystem. The economic valuation of these ecosystem services are worth many trillion dollars annually (Costanza et. al., 1997). However apart from the provisioning services, the other three types of ecosystem services often do not carry any price tags and hence societal concern are very limited over deterioration and conservation of ecosystem services.

Ehrlich & Mooney (1983) first coined the term 'ecosystem services' in order to raise awareness about the degrading habitats, functions and services of ecosystems primarily because of anthropogenic activities. Since then, there has been growing attention and research on ecosystem services from various analytical angles and over the period of time various definitions and typologies of ecosystem services have been proposed by different scientists (Daily, 1997; De Groot et al., 2002; Kremen, 2005; Boyd and Banzhaf, 2007; Fisher and Turner, 2008; Wallace, 2007).

The Millennium Ecosystem Assessment (2003) defines ecosystem services as "the benefits people obtain from ecosystems". The word ecosystem service includes both goods and services. It can also be understood as a utilization concept which is primarily anthropocentric in reference to the multitude of natural resources and processes within an ecosystem. The Millennium Ecosystem Assessment (Arico et al., 2005) has included four basic categories of ecosystem services (Table 2) that are supporting services, provisioning services, regulating services and cultural services (Gomez-Baggethun et al., 2010).

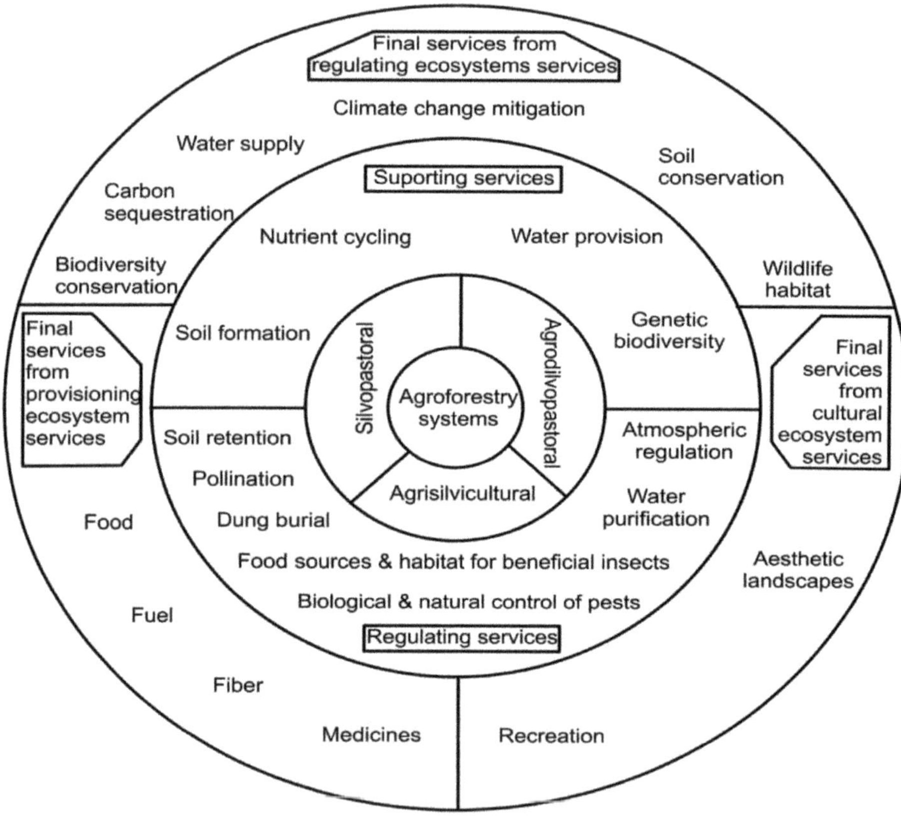


**Figure 24: Ecosystem services of agroforestry (Source: Bucheli & Bokelmann, 2017)**

## Productivity in Agroforestry Systems

One of the benefits that agroforestry systems have is the increased biomass productivity as compared to monocropping systems, primarily because of the spatial and temporal complementarities in resource-capture of trees and crops i.e. trees acquire resources in space and time that the crops alone would not. This can also be further explained on the basis of niche differentiation of both crops and trees in the same system which causes lesser competition and better coexistence (Cannell et al., 1996; Ong et al., 1996; Palma et al., 2007; Garrett and McGraw 2000; Garrity, 2004; Williams-Guille´n et al., 2008; Nair et al., 2009). Comparison of productivity of polycultures and monocropping systems can be done by Land Equivalent Ratio (LER). The value of LER greater than 1 shows that polyculture is advantageous over monocropping and less than 1 shows disadvantage. Werf et al. (2007) calculated LER in lowland England for two lowland poplar silvoarable trial systems and concluded that LER values stayed above 1 for 12 years after establishment. Similarly, biophysical modeling of hypothetical silvoarable systems in Spain, France and the Netherlands using the YieldSAFE model was done for which LER values remained between 1 and 1.4 suggesting higher productivity in case of integrated agroforestry systems than monocultures (Graves et al., 2007).

**Table 2: Ecosystem services and its components as per Millennium Ecosystem Assessment (2003)**

| S.No. | Ecosystem Service Type | Examples |
|---|---|---|
| 1 | Provisioning<br>(Physical entities/material goods) | Food, raw material, water, genetic resources, pharmaceutical resources, ornamental resources (Andrew and Masozera, 2010) |
| 2 | Regulating | Climate control and modification, air quality regulation, water purification, moderation of extreme events, maintenance of soil fertility, prevention of erosion, disease control, maintenance of soil fertility and biological control (Andrew and Masozera, 2010; Sileshi et al. 2007; Dhanya et al., 2014) |
| 3 | Supporting | Soil formation, nutrient cycling, biodiversity conservation and primary production (Andrew and Masozera, 2010) |
| 4 | Cultural | Recreational, spiritual, aesthetic, educational, tourism (Andrew and Masozera, 2010) |

The sustainable practice of agroforestry with increased biomass productivity supports the rural livelihood as it provides diverse forms of goods that increase the chances of good financial returns by providing periodic and annual revenues from rotation and reducing the risks associated with farming single commodities (Benjamin et al., 2000).

## Carbon Sequestration in Agroforestry Systems

Carbon sequestration is the removal and storage of inorganic form of carbon from atmosphere into various sinks through different physical or biological processes. Plants convert inorganic carbon as biomass which is locked up in a stable form for a time period which mainly depends on longevity of plant species. The potential of carbon sequestration by agroforestry system depends upon the system design, species composition, tree density per unit area, age of species, environmental conditions, geographic location, management practices and the end product (Jose, 2009).

Even the simplest agroforestry system has structural and functional complexity in comparison to monoculture agricultural systems. The incorporation of trees with agricultural crops increases the amount of carbon as aboveground and belowground biomass for a longer period of time than in monoculture agricultural system while still allowing the agricultural crops to grow at the same time (Sharrow and Ismail 2004; Kirby and Potvin 2007; Jose, 2009). Schroeder (1994) estimated average carbon stored by agroforestry systems to be 9 Mg C $ha^{-1}$ in semiarid, 21 Mg C $ha^{-1}$ in subhumid, 50 Mg C $ha^{-1}$ in humid and 63 Mg C $ha^{-1}$ in temperate regions. Nair et al. (2009) concluded that in a fodder bank agroforestry system of West African Sahel, the carbon sequestration potential of above and belowground vegetation was 0.29 Mg $ha^{-1}$ $yr^{-1}$ while in mixed species stands of Puerto Rico it was estimated to be 15.21 Mg $ha^{-1}$ $yr^{-1}$. Maikhuri et al. (2000) estimated average potential of agroforestry systems for carbon sequesteration to be 25tC $ha^{-1}$ over 96 million ha. Agroforestry systems hold more carbon in comparison to degraded system. In central Himalayan agroforestry system, the above ground biomass accumulation was calculated to be 3.9 t $ha^{-1}$ $yr^{-1}$ which is much higher as compared to degraded forestland with 1.1 t $ha^{-1}$ $yr^{-1}$.

Dixon (1995) estimated 585–1,215 million ha of land under agroforestry in Africa, Asia and America with total global potential (vegetation and soil) to be 1.1–2.2 Pg of carbon over 50 years. Whereas, worldwide a land area of 1,023 million ha was estimated to be under agroforestry by Nair et al. (2009) with carbon sequestration potential of 1.9 Pg of carbon over 50 years which can further be enhanced as much better sink considering the chances of large extent of degraded croplands and pasturelands to improve by agroforestry. Improving current management practices in existing agroforestry practices can result in an additional 17,000 Mg C $y^{-1}$ to be sequestered by 2040. Also, 630 million ha of unproductive croplands and grasslands if converted to agroforestry can sequester 586,000 Mg C $y^{-1}$ by 2040 (IPCC, 2000).

## Agroforestry for Increased Biodiversity

Agroforestry can be understood as planned biodiversity with the species composition chosen by farmers and hence are more diverse system than monoculture of crops. The tree species in agricultural landscapes often act as keystone species that also facilitate the movement of wildlife through a landscape that may otherwise be too hostile (Manning et al., 2009). Agroforestry systems, in some cases, do support as high as 50-80% of biodiversity of comparable natural systems (Noble and Dirzo, 1997).

The understorey botanical composition was found to change as trees matured in poplar silvopastoral systems with swards dominated by *Agrostis capillaris, Holcus lanatus* and *Poa annua* under the tree canopy while *Lolium perenne, Poa trivialis, Trifolium repens* and *Cirsium arvense* were more common in open pasture (Crowe and McAdam, 1993). Similarly higher biodiverse systems were observed in case of invertebrates (Cuthbertson and McAdam, 1996; Dennis et al., 1996), arthropods (Peng et al., 1993), some species of carabid beetles (Phillips et al., 1994), spiders

(Burgess et al., 2003) bank voles, wood mice, field voles and common shrews (Klaa et al., 2005; Wright, 1994) under different agroforestry systems as compared to monoculture.

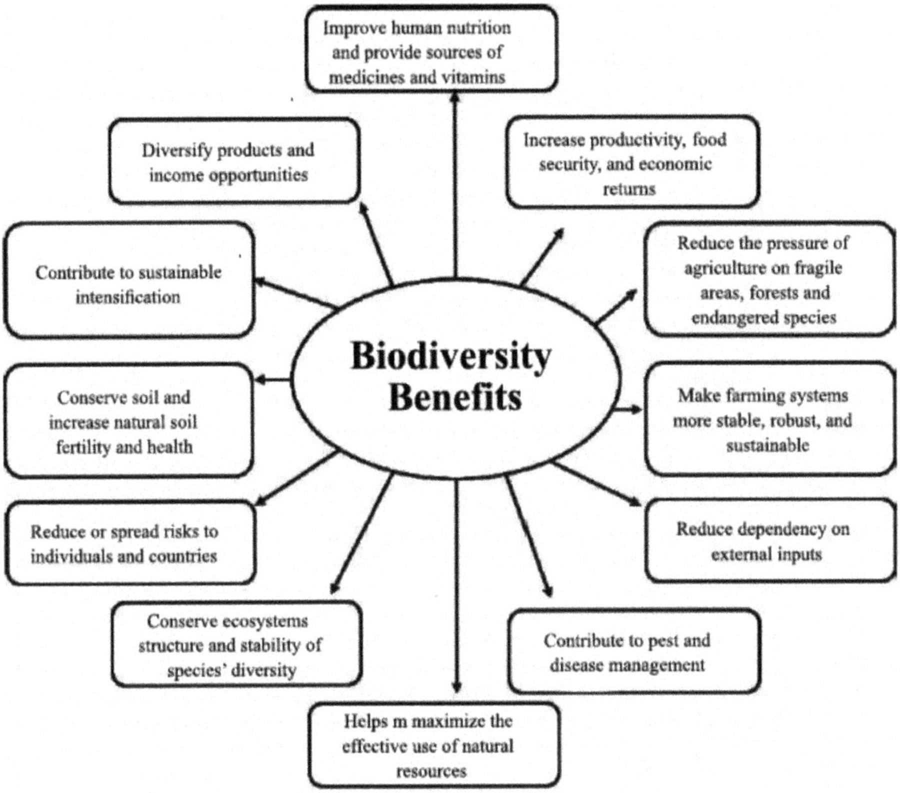


**Figure 25: The benefits provided by biodiversity (Source: Thrupp, 1997).**

According to McAdam et al. (2007) the silvopastoral systems even at an early stage of development attracted various woodland and grassland species of birds. Trees in agro-ecosystems in Rajasthan and Uttarakhand (India) supported threatened cavity nesting birds and were also found to provide food and habitat to many other species of birds (Pandey et al., 1993).

According to Jose (2009) agroforestry system enhances and conserves biodiversity by (a) providing habitats for different species; (b) preserving germplasm of sensitive species; (c) reducing the chances of natural habitat conversion and exploitation (Montagnini et al., 2011); (d) creating corridors between habitat remnants which conserve the biological entities (Montagnini et al., 2011); (e) providing ecosystem services that proliferate the survival and growth of biotic communities such as purification of water and air, soil erosion prevention, ground water recharge and nutrient cycling; (f) acting as refuge to biological communities after catastrophic events such as fire (Griffith, 2000).

Agroforestry is a well diverse sustainable practice from all point of views including social, environmental and economical benefits for land users at all levels (Watson et al., 2000). Shastri et al. (2002) in Karnataka (India) reported agroforestry systems to have more number as well as diversity of biological units than non-agroforestry system with 952 individuals belonging to 93 species in just 1.7 ha of agroecosystem. The overall trees i.e. 556 trees/ha and diversity (diversity index 3.5) was also found to be more than non-agro ecosystem with 354 trees/ha and a species diversity of 3.87.

The biological components in an agroforestry system can be planned or unplanned. In general, the agroforestry systems contain planned diversity with selected species and are more biologically diverse than corresponding monoculture crops. Structurally heterogeneous perennial vegetation, humus rich soil, better management practices and permanent litter layer are some of the conditions under agroforestry that provide appropriate habitable conditions for diverse soil fauna and microflora that may not be present in simpler and regularly disturbed agricultural systems (Thiollay 1995; Lavelle et al. 2003). Over the last two decades, several studies have successfully demonstrated that coffee agroforestry systems are important for biodiversity conservation. However this is always not the case and exception to this was observed in some traditional mixed cropping systems (Thurston et al. 1999). Similarly less biologically diverse agroforestry systems were recorded in certain shaded coffee plantations that consisted of one crop and only a handful of tree species.

Barros et al. (2008) have stated that agroforestry when compared to other forms of land use have a greater abundance and diversity of soil macro invertebrates which are also sensitive indicators of natural systems, land use and its management. Similarly greater densities and biomass of Diplopoda and Isopoda with an exception of Isopetrain comparison to adjacent secondary forest were found in the litter layer of agroforestry systems by Tapia-Coral et al. (1999). The results suggested that agroforestry systems support diverse communities of these organisms that help in different important ecological processes.

Several management practices that help in maintaining tree species richness, vertical complexity and canopies have found to support various bird species in agroforestry (Perfecto *et al,* 1996; Greenberg *et al,* 1997a, b). Comparison of bird species richness and abundance were studied in different eleven habitat types including extensive forests to intensive agricultural land uses by Petit *et al* (1999). The results of the study showed that bird communities in coffee agroforestry systems were 43% similar to that found in lowland forest fragments whereas lowest bird species had the lowest bird species richness. The results also suggested more neotropical migratory bird species in lowland forests, coffee agroforestry and residential areas (Petit *et al,* 1999).

## Soil Enrichment in Agroforestry Systems

Soil quality can be enhanced by different soil management practices such as reduced or no tillage, application of various agricultural waste for soil fertility, mulching and less or no use of chemical fertilizers. However a multipurpose agroforestry system provides diverse form of products as well as services which include nutrient cycling and soil enrichment. Soil improvement in agroforestry systems is done by (a) elevated nitrogen input in the soil, primarily by nitrogen fixing trees; (ii) increased availability of nutrients decomposition of tree biomass; and (iii) greater uptake and utilization of nutrients that may run off or leach out from deeper layers of soils by deep-rooted trees (Nair, 2011).

Semwal et al. (2003) studied chemical characteristics and decomposition patterns of six multipurpose tree species, viz., *Alnus nepalensis, Albizzia lebbek, Boehmeria rugulosa, Dalbergia sissoo, Ficus glomerata* and *F. roxburghii* in a mixed plantation in Central Himalayas. They concluded that the rate with which N and P was released during rainy season was sufficient for the requirements of Kharif crops even in case if the only source of nutrients was leaf litter.

Agroforestry systems can sustainably enhance and restore the soil productivity for longer terms. The integration of trees and crops in tropical agroforestry systems have proved to biologically fix nitrogen and enhance soil physical, chemical and biological properties by adding significant amount of above and belowground organic matter that primarily release and recycle nutrients in agroforestry systems (Nair and Latt 1997; Young 1997; Schroth and Sinclair 2003).

In the southern United States higher soil organic matter and microbial biomass was observed in alley cropping systems of pecan and cotton in comparison to monoculture cotton by Lee and Jose (2003). Similarly, Seiter et al. (1995) by using 15N injection technique in red alder and maize alley cropping system in Oregon observed that about 32-58% of total nitrogen in maize was obtained from nitrogen fixing red alder. The transfer of nitrogen in this case was also found to be related with the distance of crops from the alder trees i.e. more nitrogen from alder trees was transferred with decreasing distance between the two. Improved soil porosity (Udawatta et al., 2008a), soil aggregates stability, soil carbon, soil nitrogen, and soil enzyme activities (Udawatta et al., 2008b) was observed in soils under agroforestry buffers as compared to monoculture crops in the Midwest Region of the United States. Nair et al. (2010) has reported that more C was stored in deeper soil layers near the tree in agroforestry systems as compared to tree less systems. The higher soil organic content was also associated with higher species richness and tree density in the system. Also in deeper soil profiles, the C3 plants (trees) were reported to contribute more C in the silt + clay fractions which result in more stable C than C4 plants.

The other major aspect of soil improvement under agroforestry system is through soil conservation. According to UNEP (2004) in developing nations, about 1.9 billion ha of land, a third of total farmland is estimated to be degraded through erosion, salinity and loss of fertility. Agroforestry systems have the potential to ameliorate these degraded lands by reducing erosion and desertification for rehabilitation by conserving soil and water.

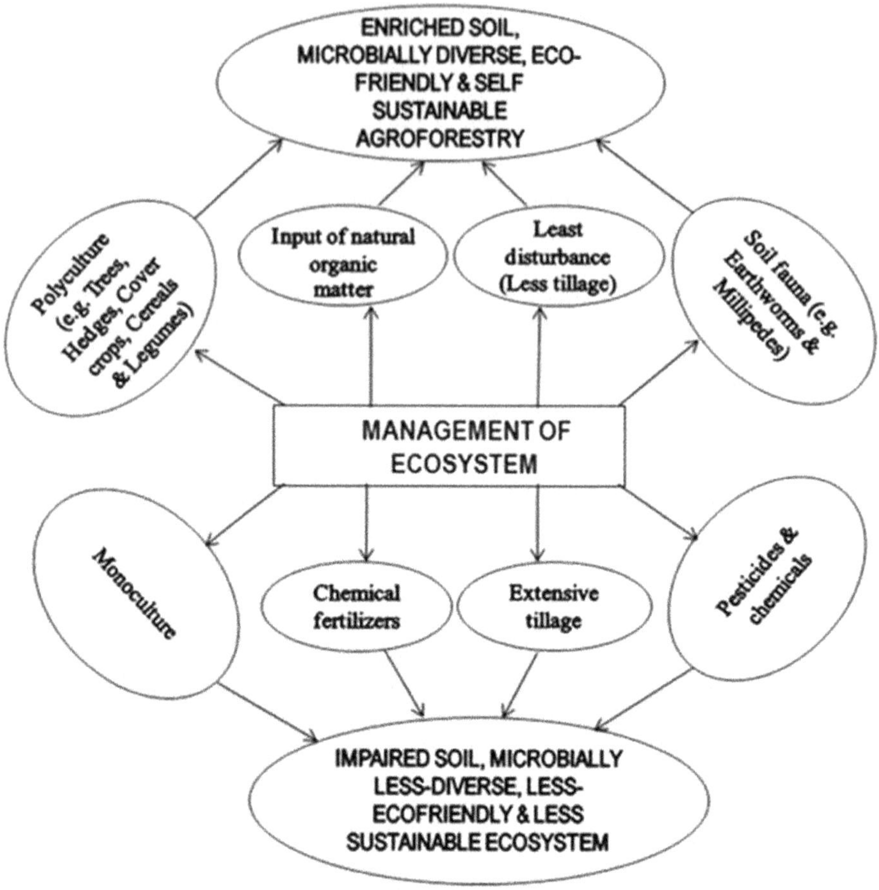


**Figure 26: Microbial Biodiversity in Agroforestry Systems (Source: Sridhar and Bagyaraj, 2017)**

## Improved Air and Water Quality in Agroforestry Systems

Agroforestry practices such as windbreaks and shelterbelts reduce wind chills and velocity, protect livestock and crops, increase biodiversity, act as sink for carbon sequestration, limit particulate matters in air, reduce noise in air and mitigate odor from concentrated livestock operations (Tyndall and Colletti, 2007). In conventional agricultural systems only about less than half of the applied fertilizer is taken up by crops. The excess of the chemical fertilizers that are not taken up by crops usually end up either in surface runoff or leached into the sub-surface water supplies where they cause contamination of receiving water bodies (Cassman, 1999). These excess fertilizers are the cause of serious environmental problems such as eutrophication and deterioration of water and biotic health in large part of the world. Agroforestry system such as riparian buffers can reduce non-point source pollution from agricultural fields. These systems clean the surface run-off as well as subsurface pollution by reducing the runoff velocity which increases infiltration and uptake/ retention of excess nutrients (Udawatta et al. 2002; Lee et al. 2003; Anderson et al. 2009). The excess of nutrients are captured by the tree roots from deeper soil horizons which remain retained in plant-soil system thereby enhancing its storage and reducing dependency on other sources for nutrient requirement (Michel et al., 2007).

Trees with deep roots are like safety nets that capture the excess of nutrients that leach below the rooting zone of agronomic crops. This is primarily done by the process of dendroremediation which enable the uptake of nutrients, pesticides and other harmful organic and inorganic chemicals via roots of the plants where they are either metabolized or stored into less harmful form. These nutrients are again supplied to the plant-soil system through root turnover and litterfall (van Noordwijk et al. 1996; Allen et al. 2004; Nair and Graetz, 2004).

Trees also have a longer growing season than most agronomic crops, which increases nutrient use and use efficiency in an agroforestry system by capturing nutrients before and after the cropping season. Allen et al. (2004) in a pecan–cotton alley cropping system in northwest Florida reported at a depth of 0.9 m about 72% reduction in nitrate-N as compared to monoculture cotton system. Similarly, enhanced soil nutrient retention and reduced nutrient transport in surface and subsurface water was observed by Nair et al. (2007) in silvopastoral system.

## Summary

There is no universally accepted typology of ecosystem services. However in simple words it can be described as an anthropocentric concept with the benefits that are derived from ecosystems. Agroforestry is the integration of trees and crops grown on farms and other rural areas for regeneration of soil, better productivity, livestock sustainability, fuel, shelter, medicines and minor products. Agroforestry systems provide provisioning, supporting, regulating and cultural services which are much pronounced at local, regional and global level in comparison to mono agriculture systems. Incorporation of trees and crops works as a multifunctional land-use solution to various global environmental issues and problems. It provides better rural livelihoods in terms of economic commodities and alleviates resource use pressure on conservation areas.

# Chapter 9

# Soil Health Improvement in Agroforestry

*Manish Kumar**, *S. Sarvade, Mukesh Topwal and Priyanka Chandra*

## Introduction

Agroforestry is sustainable land use system to practice agriculture without deteriorating agro ecosystem services while maintaining or improving productivity and maintaining stability. It is a successful and practical alternative to handle land degradation and to achieve eco-restoration and sustenance of soil resources under fragile agro-ecosystems. Agroforestry's potential to improve soil quality has been widely recognized since its beginning. It has been long since agroforestry has proved soil health benefits, but such practices remain to be fully accepted by the agriculture community and researchers. But recently, a renewed interest is increasing on soil quality and soil health as indicators of agricultural sustainability. The agroforestry can improve the major measurable soil metrics that define soil health on the evidence of 28 thematic research papers (Dollinger and Jose. 2018).

Healthy soil is the most critical resources for the health of natural and agro ecosystems so that they can maintain food production as well as provision of ecosystem services. Although the term 'soil health and soil quality' are often used synonymously, Doran et al. (1996) characterized the soil health as "the continued capacity of soil to function as a imperative living system, within ecosystem and land-use boundaries to sustain biological productivity, maintain the quality of air and water environments, and promote plant, animal, and human health." According to the USDA, soil health is "the continued capacity of the soil to function as a vital living ecosystem that sustains plants, animals and humans" (USDA-NRCS 2018). Soil quality has been defined by the Soil Science Society of America Ad Hoc Committee on soil quality as "the capacity of a specific kind of soil to function, within natural or managed ecosystem boundaries, to sustain plant and animal productivity, maintain or enhance water and air quality, and support human health and habitation" (Karlen et al. 1997). Soil quality is used with regards to both natural and managed ecosystems, but soil health is primarily used with regards to managed ecosystems, principally agroecosystems. In general, soil health refers to broader aspects and multiple soil functions, and puts particular emphasis on soil as an integrated, dynamic, living system. Moreover, soil health and soil quality terminologies were widely used in the studies related to the soil property that governs the productivity of different terrestrial systems (Katyal *et al.*, 2016). Soil quality can be used to describe more complex soil characteristics such as soil organic matter, nutrient status, soil structure, etc. Soil health is a key indicator of the condition of natural capital, and is considered as an integrative property that shows the capacity of soil to respond to agricultural management by maintaining both agricultural production and provision of ecosystem services (Kibblewhite *et al.*, 2008). Agroforestry and other management practices influence these attributes and have the capability to enhance or diminish soil health. Soil properties (chemical, physical and biological) were significantly improved under poplar based agroforestry system and tree density of 500 and 1000 trees/ha proved to be the best tree density for improving soil health (Uthappa *et al.*, 2015). Many researchers were emphasized on the indicators of soil health and reported the soil organic matter, respiration, microbial biomass (total bacteria and fungi,) and mineralizable nitrogen, soil pH, EC, CEC, available soil N, P, K, soil porosity, water holding capacity, structure and tilth etc. Brevik (2009) and Katyal *et al.*, (2016) enlisted the commonly used indicators for studying soil health (Fig.1).

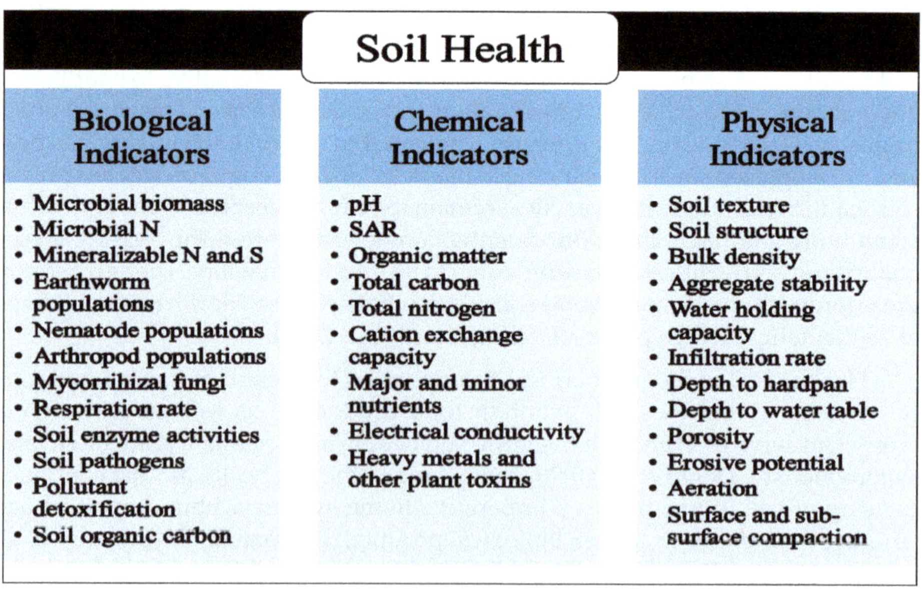


**Figure 27. Indicators of soil health (Source: Brevik, 2009; Katyal *et al.*, 2016)**

Chemical, physical and biological are three main categories of soil indicators which regulates soil health and its functions. Chemical indicators provide the information about the equilibrium between soil solution (soil water and nutrients) and exchange sites (clay particles, organic matter); the nutritional requirements of plant and animals and their availability to plants and animals. Physical indicators can give us information about soil hydrologic characteristics, such as water entry and retention that influences availability to plants. Biological indicators can tell us about the organisms that are responsible for decomposition of organic matter and nutrient cycling. The relationship between indicator type and soil functions is an important factor which decides the productivity and sustainability. Organic matter has the most widely recognized indicator of soil quality. Organic matter is interconnected to all soil functions. It affects other indicators, such as aggregate stability, water retention (physical), nutrient retention and availability (chemical), and nutrient cycling (biological). The detail soil parameters indicative to chemical, physical and biological indicator of soil health and related function was depicted in Table.3.

**Table 3. Soil health indicators and related functions (Adapted from Doran and Parkin, 1994; Larsen and Pierce, 1994)**

| Soil indicators | Relationship to soil condition and functions |
|---|---|
| I. Physical | Physical Stability & Support, Water Relations, Habitat |
| Soil texture | Retention, transport of water & chemicals, soil erosion |
| Soil depth, top soil, rooting | Estimation of productivity potential and erosion |
| Infiltration and B.D. | Potential for leaching, productivity and erosivity |
| Water holding capacity | Available water, water retention, transport and erosivity (Available water depends on BD, texture and SOM) |
| II. Chemical | Nutrient Cycling, Water Relations, Buffering capacity |
| SOM (total OC and N) | Soil fertility, stability, erosion, site normalization |
| pH | Biological and chemical activity threshold |
| EC | Plant and microbial activity threshold |
| Available N,P,K | Plant available nutrient, potential productivity and environmental quality indicators |
| III. Biological | Biodiversity, Nutrient Cycling |
| Microbial biomass C &N | Microbial catalytic potential and repository for C & N |
| Mineralizable N | Soil productivity and N-supply |
| Soil respiration, water content and temperature | Microbial activity measures, process modelling |

## Distribution, Extent and Effects of Soil Degradation

In India, the estimated area under wasteland is approximately 30 to 175 Mha due to environmental degradation by way of soil erosion by water and wind, soil salinization/ alkalization/acidification, water logging etc. Only intensification is not the main driver of land degradation, but rather the policy environment that empowered improper land utilize and injudicious input use, particularly excessive irrigation. The extent and severity of such degradation, however, varies. Maintenance of soil health and/or curbing its degradation is of paramount importance now a day. If its health is maintained, a large number of global challenges *viz.*, food security, water security, waste recycling, protection of biodiversity and climate change etc. can be resolved. In fact, maintenance of soil health is ultimately linked to the plant, animals and human health.

## Reasons of Soil Health Degradation (Source: Katyal *et al.,* 2016)

- Population pressure
- High deforestation
- Change in land use pattern
- Intensive soil farming operation
- Use of high yielding varieties
- Irrational use of fertilizers, pesticides and insecticides
- Faulty Irrigation management practices
- Switching over to fossil fuel energy
- Modern intensive soil management practices and extension

In India, about 90 Mha land is affected by various types of soil physical constraints (Table. 4) and 6.73 million ha by chemical constraints such as salt-affected soils. These soils (both irrigated and rainfed) produce very low crop yields and have lower fertilizer nutrient use efficiency. Mass flow and diffusion are the two major physical processes by which nutrient ions are transported to the plant root; immobile nutrients are also absorbed through root interception. Management practices which alter water availability and root growth in time and space are likely to influence these physical processes. The major soil physical constraints in Indian soils include low water retention and high transmission, slow permeability, surface and sub-surface mechanical impedance and shallow soil-depth. It either restricts crop growth or reduces efficiency of basic inputs *viz.,* water and fertilizer nutrients (Chaudhari, 2016).

**Table 4. Area and states affected by various soil physical constraints in India (Source: Chaudhari, 2016)**

| Physical constraints | Area (Mha) | Main states affected |
|---|---|---|
| Shallow depth | 26.40 | Andhra Pradesh, Maharashtra, West Bengal, Kerala and Gujarat |
| Soil hardening | 21.57 | Andhra Pradesh, Maharashtra and Bihar |
| High permeability | 13.75 | Rajasthan, West Bengal, Gujarat, Punjab and Tamil Nadu |
| Subsurface hard pan | 11.31 | Maharashtra, Punjab, Bihar, Rajasthan, West Bengal and Tamil Nadu |
| Surface crusting | 10.25 | Haryana, Punjab, West Bengal, Orissa and Gujarat |
| Temporary water logging | 6.24 | Madhya Pradesh, Maharashtra, Punjab, Gujarat, Kerala and Orissa |

## Remediation of Salt Affected Soil through Agroforestry

Salt-affected soils have been adversely altered the growth of most plants by the action of soluble salts (saline soils), the exchangeable sodium (Na) percentage (sodic/alkali soils) and/or both (saline-sodic soils). Salinity reduces water availability for plant use. The high salinity levels hinder water absorption, inducing physiological drought in the plant. The soils may have sufficient water, however plant roots are unable to assimilate the water due to unfavourable osmotic pressure. This is known as the osmotic or water-deficit effect of salinity. Generally, plants are mostly sensitive to salinity during germination and early growth and development. The second impact of salinity is observed when excessive amounts of salt enter the plant through transpiration stream and injure leaf cells, which further reduces growth and development. This is called the salt-specific or ion-excess effect of salinity. The symptoms may include restricted root growth, marginal or leaf tip burning/ browning, inhibited flowering, reduced vigor and reduced crop yields. Sodic soils contain sufficient exchangeable $Na^+$ (Na-occupied negatively charged on silt and clay) which adversely affect soil physical properties, particularly soil structure, infiltration and aeration, and therefore plant growth. A compact calcichorizon ($CaCO_3$ concretions or kankars), often present in sodic soils, can restrict root penetration and movement of water and air within the soil profile.

Agroforestry has been shown to be a low cost and highly remunerative option that can be implemented by resource-poor farmers with the potential of increasing their income and diversifying farming enterprises on salt affected soils. Therefore in the view of biological amelioration of salt-affected soil, planting trees along with natural vegetation such as grasses is necessary. Trees have some inbuilt mechanism for tolerance a certain level of salinity. Appropriate planting techniques have been standardized for raising tree plantations in saline soils (sub-surface planting, ridge-trench method, furrow irrigation system) and sodic soils (ridge-trench method, auger-hole technique, furrow method). Growing of grasses in combination with *Prosopis juliflora* and *Acacia nilotica* improved the soil health. Ten years after planting, silvipastoral system (*Prosopis juliflora+ Leptocloa fusca* for four years followed by Berseem (*Trifolium alexandrium)* for six years produced significantly more fodder than other system (Singh *et al.*, 2014). This might have been because of improvement in soil health due to the combined effect of *P. juliflora* and *T. alexandrium* nitrogen fixing grass species. Kumar et al. (1990–1991) found that *T. alexandrium* yield in salt-affected soils was less affected by tree canopy species than were other fodder crops. Among the various tree species tested, *Prosopis juliflora, Acacia nilotica, Tamarix articula, Casuarina and Eucalyptus* etc. showed better ability to ameliorate salt affected soils. The tree species for reclamation of salt affected soil are listed in Table 5 and 6.

**Table 5. Example of tree species as per tree classification based on nature of tolerate/sensitive to different soils with varying pH (Tomar and Gupta, 1984-94)**

| pH | Agroforestry tree species |
|---|---|
| >10.0 | *Prosopis juliflora, Acacia nilotica, Tamarix articulate, Casuarina equisetifolia* |
| 9.5-10.0 | *Acacia catechu, Albizia lebbeck, Butea monosperma, Cordiarothii, Eucalyptus tereticornis, Pithecellobium dulce,, Pongamia pinnata, Tamarindus indica,, Terminalia arjuna* |
| 9.0-9.5 | *Acacia auriculiformis, Albizia procera, Azadirachta indica, Bambusa spp., Cassia siamea, Dalbergia sissoo, Grevillea robusta, Kigelia pinnata, Leucaena leucocephala, Melia azedarach, Moringa oleifera, Morus alba, Sesbania sesban* |

| <9.0 | *Bauhinia variegata, Bombaxceiba, Cedrelatoona, Ficusvirens, Hardwickiabinata, Populus deltoids, Tectona grandis, Thespesia populnea* |
|---|---|

**Table 6. Example of tree species in different tree classification on the basis of degree of tolerate to sensitive to soil with varying EC (Tomar and Gupta, 1984-94)**

| EC | Agroforestry trees species |
|---|---|
| Tolerant (EC 25-35 dS/m) | *Acacia farnesiana, Parkinsonia aculeanata, Pithecellobium dulce, Prosopis juliflora,Tamarix articulata, Tamarix troupii* |
| Moderately tolerant (EC 15-25 dS/m) | *Acacia pennatula, Acacia tortilis, Acacia nilotica, Casuarinaequisetifolia, Casuarinaobesa, Casuarinaglauca, Callistemon lanceolatus, Eucalyptus camaldulensis, Leucaena leucocephala* |
| Moderately sensitive (EC 10-15 dS/m) | *Acacia auriculiformis, Albizia caribaea, Casuarina cunninghamiana, Eucalyptus tereticornis, Leucaena shannoni, Pongamia pinnata, Samanea saman, Terminalia arjuna* |
| Sensitive (EC 7-10 dS/m) | *Acacia deamii, Albizia guachapele, Caesalpinia eriostachya, Caesalpinia velutina, Haematoxylon brasiletto, Salix spp, Syzygium cumini, Syzygium fruticosum, Tamarindus indica* |

## Soil Health Improvement through Agroforestry Intervention and Practices

Since land resources are finite, requisite measures are required to reclaim degraded and wastelands, so that these areas need to be brought under utilization by using reforestation of potential species. Agroforestry has vast potential to improve soil health by the process like addition of leaf litter from perennial leguminous tree species, decomposition and nutrient release through microbial activities which helps to improve soil physical, chemical and biological properties. The biomass addition through dead root systems of the tree species leads to improvement in soil organic carbon that improves soil health indicators. Soil biological diversity was considered to be vital for improving soil health properties. Their role in decomposition and degradation of soil OM and nutrient cycling improves soil chemical and physical properties, which improves the soil fertility. Sisay and Ketema (2015) studied abundance and diversity of soil macro-fauna under agroforestry and maize crop and concluded that the abundance and diversity of soil invertebrate macrofauna was positively influenced under indigenous agroforestry (Shannon diversity index= 1.28) as compared to cultivation of maize as sole crop (Shannon diversity index= 1.17). *Acacia senegal* tree had positive influence on soil microbial biomass and inorganic-N content and the highest values were recorded at the surface soil and at 0-25 cm layer (Fall *et al.*, 2012). Kumar *et al.*, (2008) revealed that the highest annual concentration of microbial biomass carbon under *Terminalia arjuna* followed by *Eucalyptus tereticornis* and *S. mahagoni*, whereas the highest microbial quotient was observed under *Pongamia pinnata*. The enzyme activities are greater in agroforestry practices due to differences in litter quantity and quality and root exudates (Mungai *et al.*, 2005). Dobo *et al.* (2018) compared the arbuscular mycorrhizal fungal diversity and spore density between nine agroforestry systems based either on Cordia, *Millettia* and *Erythrina* associated with coffee-cultivation and observed significant effects on AMF spore abundance among different combinations of tree-enset-coffee agro-forestry practices. However, *Millettia ferruginea* and *Erythrina brucei* land uses had higher spore numbers than with *Cordia africana* as a shade tree in agroforestry.

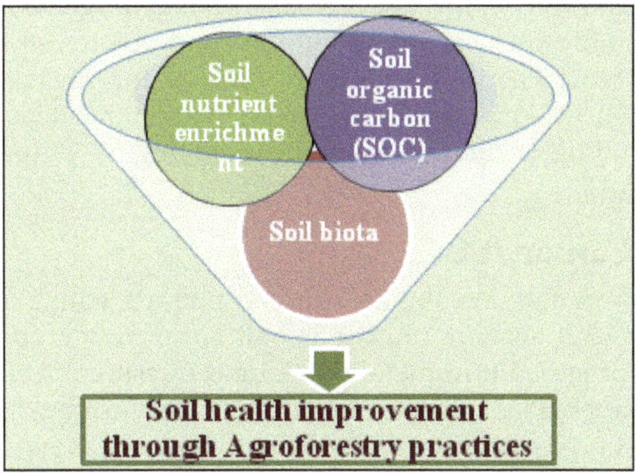


**Figure 28: Soil health improvement mechanism through agroforestry practices**

Energy requirement of soil microbes was fulfilled through addition of adequate amount of crop residue and leaf litter. It leads to improve soil physico-bio-chemical properties through decomposition of soil organic matter. Rousseau *et al.,* (2012) studied a minimum set of four well-accepted abiotic soil quality indicators (bulk density, sum of bases, pH and carbon) which was able to separate cacao based agroforestry and forests into five distinct clusters along a low-to-high "soil quality" gradient. The cacao AFS plots were rated as "good" soil quality and not differ from the forest systems. The cacao-based AFS were able to conserve soil and provide a high level of soil-related ecological services. Sarvade *et al.,* (2014a) studied soil properties under agri-silviculture systems of Terai Region, Northern India and observed that 1.73% organic carbon under *Populus deltoids* based agri-silviculture system as compared to sole cropping of wheat field (0.60%). They found the highest available soil $N_2$, $P_2O_5$ and $K_2O$ content (200.42, 18.67 and 169.12 kg ha$^{-1}$, respectively) in agri-silviculture systems having nitrogen fixing tree species such as *Leucaena leucocephala*. Bisht *et al.,* (2015) studied the impact of different tree spacing treatments on soil properties and revealed that organic carbon (0.47%) and N (199.8 kg ha$^{-1}$), P (15.6 kg ha$^{-1}$) and K (226.4 kg ha$^{-1}$) contents were recorded maximum under 3 m × 3 m spacing of *Eucalyptus* based agroforestry system. Fall *et al.* (2012) studied an impact of distance and depth on microbial biomass and mineral nitrogen content under *Acacia Senegal*. They observed the negative correlation of microbial biomass and inorganic-N content with distance from tree stem and the depth and the highest values were recorded at the foot tree (R0) and at 0-25 cm layer. The inorganic-N was mostly dominated in nitrate form ($NO^{3-}$) and ammonia form ($NH^{4+}$) during dry season and wet season respectively. Zhang *et al.,* (2018) investigated, by means of molecular techniques, how plant-tree associations in agroforestry systems shape the fungal community and observed that out of three systems *viz.* barley with *Populus euramevicana* or *Taxodium distichum* and barley only, greater fungal diversity was found in the rhizosphere compared to bulk soil. However, higher abundance of certain rhizospheric fungi was observed in *P. euramevicana* based systems than in *T. distichum* based systems. Soil biological properties were significantly influenced both by tree densities and soil depths. The activities of dehydrogenase and phosphatase enzyme were significantly higher under 1000 and 500 trees/ha, respectively as compared to sole crop, whereas both activities decreased with increase in soil depth. Tree density of 500-1000 trees/ha proved to be the best tree density for improving soil health depending on the trade-off between tangible outputs in the system (Uthappa *et al.*, 2015).

## Soil Organic Carbon (SOC)

The organic matters are the carbon compounds which are the source of energy. These organic matters present in soil are mainly crops/plants residues. The ecosystems of agriculture mainly include the residual leaf, stem, and roots remained after harvest. Plant residues are divided into two types that are assimilated into the soil. First type is the residues which are present above the ground, these includes falling leaves and stem debris and other plant that are present above the soil surface. The second type of the residues is the root systems. These root system residues contain root exudates which are the good source of several nutrients and act as substrate carbon input for the soil. Organic matter may be directly or indirectly affect the plant growth. These act as a natural substrate for soil microbes which provides nutrition to plants indirectly through their activity. Organic matter is essential for the formation of soil aggregates and hence soil structure which ultimately determines the soil aeration and rooting habit of plants. Organic matter helps in the conservation of soil nutrients by preventing erosion and surface run off of nutrients. The addition of biomass to soil from tree legumes such as *Gliricidia sepium, L. leucocephala* through branch prunings, litter and root turnover in alley cropping and planted tree fallows significantly contribute to increased light fraction SOM. Nevertheless, the relative contribution to light fraction SOM varies significantly amongst tree species. For example, the contribution of *Sesbaniasesban* to the light fraction SOM was five times greater than species such as *C. calothrysus, Flemingia macrophylla, G. sepium, L. leucocephala,* and *S. siamea* (Barrios *et al.*, 1997).

## How do Agroforestry Tree Influences Trees-soil Biota Interactions?

Minerals and nutrients present in soil are sustained in due to the biological activities carried out by microbes and plants. Hence, the role of soil biodiversity in maintaining soil biological properties along with physical and chemical characteristics is very critical for healthy soil. The SOM determines the microbial diversity that varies with depth. Biological parameters such as microbial biomass, soil respiration, enzyme activities and microbial diversity indicate soil health. Significance of soil biodiversity for sustainable agriculture is very important as microbial diversity is the indicator for the soil health. The variation in microbial population or activities is the signs of soil degradation or amelioration. Availability of water and nutrient particularly N and Pin the soil directly correlates the plant growth. The above ground vegetation like trees is the primary source of carbon for the microbes present in the rhizospheric soil which in turn, provides nutrients to the plants. Hence, the vegetation also influences microbial community structure in the soil and their properties.

Agroforestry is also a practice that offers great promise to improve soil biological properties and its health. Agroforestry, as a sustainable land management practice, has shown the evidence of its role in improving soil quality; it could improve soil nutrient availability and soil fertility due to the presence of trees in the system and enhance soil microbial dynamics which would positively influence soil health and productivity. Several researchers are focusing their interests into agroforestry as a viable strategy to restore and sustain soil health as they nurtures the soil microbes. Soil microorganisms plays a key role in nutrient cycling, organic matter transformation, and also in improving physical characteristics of soil. The microbial biomass includes both primary and secondary decomposers, is vital for the cycling of nutrients in soil as it governs the breakdown of OM and the availability of nutrients, particularly N-mineralization. Microbes under agroforestry systems mostly concentrate to the soil surface and the dense network of fine roots of trees with mycorrhizal association increases the availability of nutrients to the under story crops.

## Role of Microbes in Soil Biological Processes

There is a strong relationship between the soil biological communities and the biological processes. Soil microorganisms are grouped into four functional assemblages: 1) decomposers, 2) nutrient transformers, 3) ecosystem engineers, and 4) biocontrollers, each composed of several functional groups. The four aggregated ecosystem functions that include carbon transformations, nutrient cycling, soil structure maintenance and population regulation can be regulated by the functional attributes of these assemblages. The decomposition of organic matter occurs in which C in litter and other organic inputs are transformed through the consecutive fragmentation and enzymatic activity of diverse decomposer organisms. The decomposition of organic matter results in the release of $CO_2$ and the synthesis of soil organic matter (SOM). While strongly linked to decomposition, the cycling of nutrients is largely mediated by soil microorganisms. The maintenance of soil structure is fostered by the combined action of plant roots and soil organisms known as "soil ecosystem engineers" that continuously changing soil physical properties and making microhabitats for other soil life forms by modify the soil by forming "biological" aggregates, pores, and channels. The biological control of pest and diseases happens through the activity of a wide range of soil organisms that regulate the populations of soil-borne diseases and pests to a great extent through competition, predation, and parasitism.

## Carbon Transformations and Nutrient Cycling

Soil microbes transform carbon between different ecosystems. They utilise different organic and inorganic forms of carbon as carbon and energy sources. However, the C cycle does not operate independently; it is closely coupled with that of other essential elements for microbial metabolism. This linkage happens either by means of the utilization of other elements as electron donors and acceptors (e.g. N species ranging from the most reduced, $NH^{4+}$ , to the most oxidised, $NO^{3-}$) in energy transduction or through their immobilisation and mineralisation as a feature of numerous essential element-containing biomolecules (e.g. proteins, DNA). Henceforth the accessibility of other key elements essential for life, particularly N and P, and other environmental factors such as pH, soil texture and mineralogy, temperature and soil water content control the rate of microbes consume and respire carbon. This interaction between environmental conditions and biological processes, including primary production, is the chief in controlling the unequal distribution of organic matter across the world's soils.

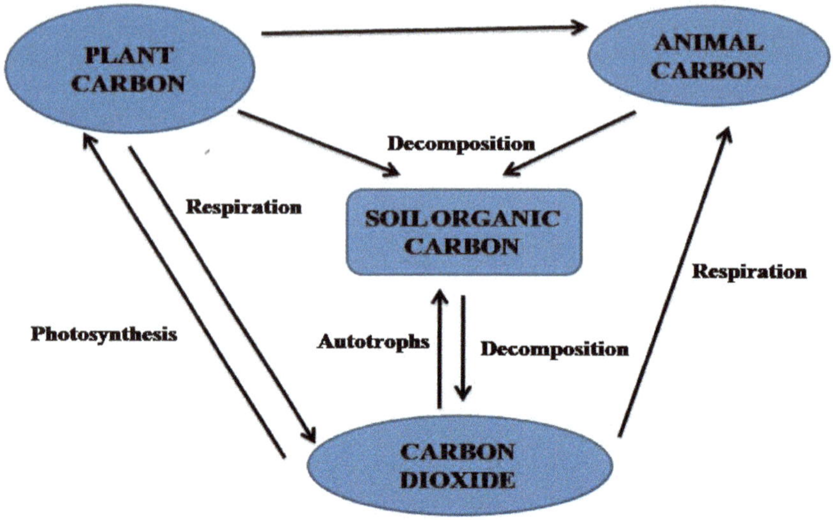


**Figure 29: Soil organic carbon fixation in agroforestry**

A major contribution of trees to soil based ecosystem services occurs as a result of aboveground and belowground organic inputs that provide C-substrates and nutrients needed for the soil organisms involved in C transformations and nutrient cycling. Carbon transformations happen amid the deterioration of organic inputs because of the aggregate activity of decomposer organisms that fragment organic inputs (e.g. earthworms, millipedes, termites and mites). This change in turn facilitates the enzymatic action by fungi and bacteria that results in the release of nutrients to the soil matrix, loss of carbon to the atmosphere as $CO_2$, and the amalgamation of SOM (Barrios 2007). Carbon transformations and nutrient cycling take place through coordinated interaction of decomposers and nutrient transformers (Kibblewhite *et al.*, 2008) and are treated here as a functional continuum. Increased nutrient availability in agroforestry systems is often associated with higher levels of SOM under trees than away from trees (Buresh and Tian 1998). Nevertheless, while increases in total SOM are closely related with increases in soil water accessibility, this is not the situation for soil nutrient availability because nutrient release is dependent on the biologically active portion of SOM (i.e. microbial biomass, light fraction SOM). The plant-soil biota interactions strongly regulate the relative contribution of organic inputs to nutrient and $CO_2$ release through mineralization processes and SOM synthesis. The quality of organic inputs impacts the decomposer biota structure and composition, thus regulates the magnitude and rate of nutrient release (Wardle *et al.* 2004). Organic resource quality is an indicator of chemical composition and has been operationally defined by the concentrations of total N, lignin, and soluble polyphenols. Organic inputs of "high quality" having low lignin + polyphenol/N ratio, will decompose quicker than those of "low quality" (e.g. high lignin + polyphenol/N ratio) and hence contribute moderately more to soil nutrient availability than to SOM formation. The woody biomass also acts as a substrate for the synthesis of SOM. Regular organic inputs through leaf litter, tree pruning, and root turnover will have long term impacts on soil carbon and nutrient stocks and thus agro-ecosystem sustainability.

## Symbiotic Interactions and Nutrient Cycling

The processes in the nitrogen cycle are to transform nitrogen from one form to another. It includes different biological processes like N-fixation, ammonification, nitrification, and denitrification. The conversion of atmospheric nitrogen gas ($N_2$) to nitrates and nitrites,by industrial and biological processes is known as nitrogen fixation. Biologically atmospheric nitrogen is fixed into an available form to be taken up by plants is carried out by free-living or symbiotic bacteria called as diazotrophs. The nitrogen fixation process is carried out nitrogenase enzyme present in the nitrogen fixing microbes that convert nitrogen to ammonia which was further converted to organic compounds in microbial cells. Microbes provide ammonia to plants in exchange of carbohydrates and other nutrients. *Azotobacter* is an example of nitrogen fixing free-living bacteria while *Rhizobium* is asymbiotic nitrogen-fixing bacteria. They resides in the root nodules of legumes plants and form a mutual association with the plant.

Agroforestry play an important role in the process of nutrient cycling and uptake. They improve the SOM which further leads to improved nutrient cycling and high soil productivity. Trees improves the supply of nutrients in the root zone of crops by biological $N_2$ fixation, by absorbing nutrients from below the root zone of the crops and reducing nutrient losses by hindering the leaching and soil-erosion. The root binds the soil particles which check the soil erosion and consequently soil nutrient. The deep-rooted trees have the ability to capture nutrient from sub-soil layers when trees have a high demand for nutrients and high amounts of plant-available nutrient are present in the subsoil. Panwar and Gupta (2015) conducted a study on litter fall, nutrient return and soil fertility under *Celtis australis* reported that the order of nutrient input from different components of litter was highest from leaves followed by miscellaneous litter and twigs. Nutrient concentration in the litter component is also an imperative aspect as it impacts on the rate of disintegration or decomposition of litter and also the quantity of nutrients released at the time of decomposition process. *Arbuscular mycorrhizal* fungi (AMF) associated with trees can complement the nutrient capture function of deep roots by increasing the recovery of nutrients from the subsoil. They may also reduce nutrient loss through leaching and associated pollution, while also increasing uptake of less mobile nutrients like phosphorus. For example, the remarkable ability of *T. Diversifolia* to accumulate large quantities of P in its biomass, as well as all other nutrients, seems to be related to the unusual specificity of its mycorrhizal associations (Sharrock *et al.*, 2004).

## Soil Structure Maintenance

Soil aggregates are the arrangement of soil primary particles and soil organic matter that are bound together by organic and inorganic compounds which forms the structural units of the soil. Hence it can be said that soil structure is a dynamic property which reflects the balance between aggregate forming factors. The formation of "biological aggregates" and their stabilization is the result of the activity of fungi, bacteria, plant roots, and macro-fauna in the soil. AM-fungi have also been considered to be the important factor for their contribution to the formation and maintenance of soil structure through hyphal networks of soil aggregates and deposition of glomalin, an AMF-specific glycoprotein strongly linked to water stable aggregation (Rillig, 2004). The role of trees is vital for promoting biological activity which contributes to soil structure maintenance in slash and mulch agroforestry practice and the comparative study of the impact of slash and mulch with slash

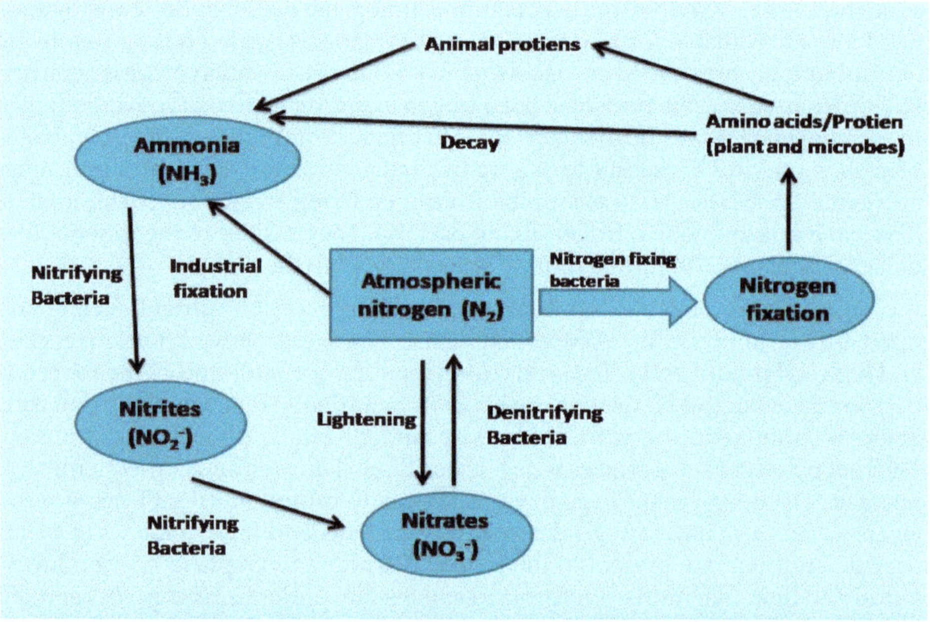


**Figure 30: Nitrogen cycle by leguminous tree in agroforestry**

and burn showed that the mean soil erosion rate after three years was about six times higher with burning than mulching (Castro *et al.*, 2009). Additionally, higher soil mesoporosity (30%) and macroporosity (19%) for mulching versus burning are consistent with increases in biological activity of various soil organisms that generate pores and channels of different sizes and shapes. These porosity differences significantly influence the available soil water storage capacity, water infiltration, surface runoff and soil erosion. When soil aggregates break into smaller pieces after wetting, erosion rates increase, and SOM is promptly exposed to microbial activity that results in C loss to the atmosphere (Barrios 2007). The dynamics of physical assurance of SOM in soil aggregates has gotten considerable attention because of its significance for soil carbon sequestration. Therefore, the overall potential effect of agroforestry on soil erosion control and soil C sequestration are clearly underpinned by tree–soil biota interactions. They foster increases in the proportion of soil aggregates that are stable upon wetting during rainfall events, and with the magnitude and diversity of soil porosity that allows a balance between infiltration and soil water storage for plants and soil organisms. These outcomes highlight the vital opportunities for the design of agroforestry practices that incorporate tree species diversity and mulch management to promote soil biological diversity and their activity optimizes the aggregate dynamics required for erosion control, C-sequestration, supply of good quality water and ecosystem services.

## Conclusion

Healthy soil is conceivably one of the most critical resources for attaining the sustainable production of goods and services, including agroecosystems. The sustainability of soil has been threatened worldwide due to faulty management practices and there has been a renewed interest in protecting the soil for sustainable production in future. Agroforestry have the potential to addresses the challenge of optimizing crop productivity while maintaining ecosystem services via improving soil health. The adoption and promotion of agroforestry should include multipurpose tree species (MPTs) to enhance the sustained provision of soil-based ecosystem services. Overall, agroforestry has the ability to (1) enhance soil organic carbon better than solo cropping system, (2) improve the soil structure and soil fertility status, and (3) increase the soil-biota or microbial dynamics which would promote the factors influencing the soil health positively. The rhizosphere had a positive influence on soil microbial biomass and inorganic-N content which might be beneficial for crops in agroforestry in N-deficient and poor fertile land. Therefore, it is imperative that agroforestry as a sustainable land management practice should arguably increase attention in our policy discussion as viable strategy to restore and sustain soil health.

# Chapter 10

# Productivity, Biomass and Carbon Storage Potentials of Agroforestry

*Manish Kumar*, Kamini Gautam and Shivaputra Bammanahalli*

## Introduction

Tree based land use system play an important role in carbon cycling and C-sequestration in terrestrial ecosystems. Forest ecosystems approximately stored 638 Gt C, with 283 Gt C in biomass alone (Keeling and Phillips, 2007). Recently, Global warming or climate change is a very serious environmental issue worldwide caused by greenhouse gases mainly ($CO_2$) and threatening our biodiversity and declining productivity. Gradually with the increase in human population, development and urbanization demand for fuel, fodder, timber, infrastructure increased instantly leading to enormous deforestation and degradation of ecosystem, soil erosion etc. and ultimately ecological instability. Further with the rise of industrialization, fossil fuel burning, other development activities have led to increase in carbon dioxide levels from 200 ppm (Ice age) to 400 ppm (2013). Climate change is already having significant and considerable cost effects on our communities and biodiversity. Unless we take immediate action to reduce the global warming emissions through capturing and fixing greenhouse gases, these effects will keep on intensifying, grow ever more harming and progressively influence the whole ecosystem. Agroforestry land use systems have great potential to sequestered carbon (C) and stabilize the $CO_2$ levels. International organisation like United Nations Framework Convention on Climate Change (UNFCCC) recognizes the role of afforestation, reforestation, and natural regeneration of forests in increasing the C storage capacity of terrestrial ecosystems.

## Productivity of Agroforestry System

Productivity refers to the rate at which energy is added to the bodies of organisms (such as primary producers) in form of biomass. Gross productivity can be understood by the overall rate of energy capture through photosynthesis by plant, while, Net productivity is the excess of energy generated by the process. It is the difference between Gross productivity (GP) and cellular respiration (R) and can be estimated by (NP = GP-R). Tree productivity and biomass are influenced by various factors like climatic, edaphic, topographic, biotic and limiting factors. The environmental factors which limit the productivity are mainly temperature, precipitation, and soil nutrient availability. The relationship between tree productivity and biomass may reflect plant adaptation and the interaction of trees and environments, and may have significant implications in global carbon cycling, climate change, and forest management. Therefore, it is quite important to understand the relationship between tree productivity and biomass.

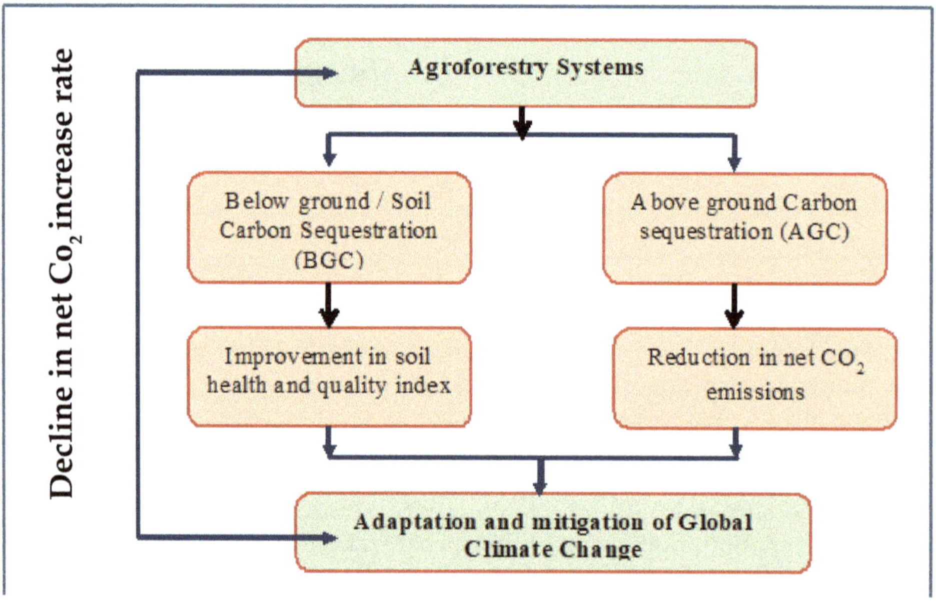


**Figure 31. Interactive approach of Agroforestry and Carbon sequestration**

There have been two general approaches to establishing a relationship between productivity and biomass. They are (1) A direct relationship between productivity and biomass is based on the assumption that productivity is the source of biomass (Keeling and Phillips, 2007). A linear regression model and nonlinear regression equation has been used to define the relationship of productivity and biomass in different forest ecosystems. (2) The relationship between tree productivity and biomass adapted from the metabolic theory of ecology. In this, the allometric relationship of rate of production (P) and biomass (M) can be described by a power function:

$P = a\,M^b$,

Where, a refer to normalization constant and b represents an allometric scaling exponent.

Plants typically capture and convert about 1.3-1.6 percent of the solar energy that reaches Earth's surface and use about a quarter of the captured energy for metabolism and maintenance. In this way, around 1 % of the solar energy reaching Earth's surface (per unit area and time) can be utilised as net primary productivity. Net primary productivity differs among ecosystems and relies upon many factors. It includes solar energy, temperature and moisture, nutrient availability, carbon dioxide levels, and community interactions. These factors influence the number of photosynthesizers is present to capture light energy and efficiently perform their duties. The distribution and abundance of plants are influenced by the biotic and abiotic factors of the environment such as temperature, moisture, light, nature of soil (texture, Organic matter, nutrients, microorganisms, pH, and amount of salt), fire, concentration of $O_2$ and $CO_2$. Plant size, which can be represented by age, DBH and height, is a most essential aspect of their biology and influence almost all other characteristics features. However, the influence of biotic factors such as tree age or size and abiotic factors such as geographic locations on the biomass and productivity not been systematically investigated.

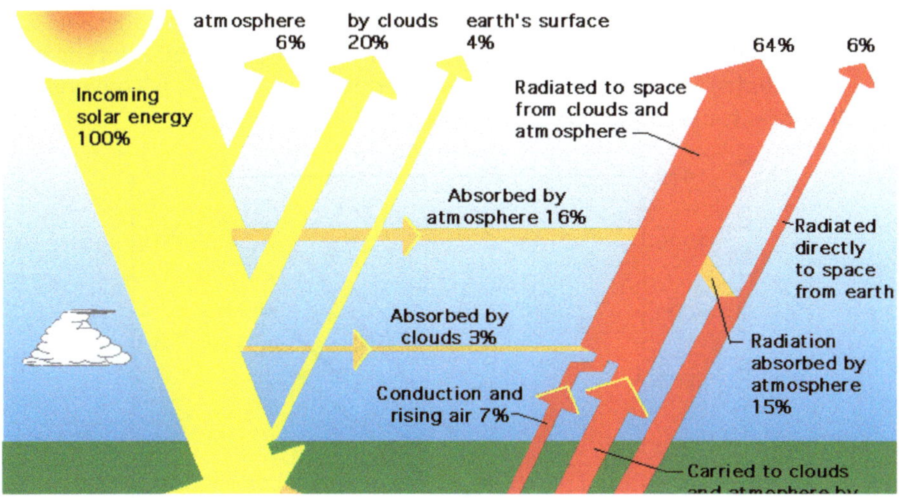


**Figure 32. Effects of clouds on the Earth's Energy Budget (image is from a NASA site)**

## Strategies for Maximizing Solar Energy Utilization, Leaf Area and Crop Growth

Photosynthesis is the biochemical process by which plant converts carbon dioxide into organic compounds, particularly sugars, utilizing the sunlight energy. Depending on carbon fixation process, the plants can be categorised into three types: C3, C4, and CAM. The C3 gramineous plants are the superior plants, which are the temperate weather crops (wheat, barley and sunflower, etc); the C4 plant is species from arid weathers or tropical weathers (e.g. Tobacco, maize, sorghum). The C3 type is generally considered less productive than C4 (Fig.2). The climatic parameters and $CO_2$ concentrations influences directly the relative abundance of C3 and C4 plants with all other factors held constant, while other variables can also influence the distribution of the plants. The rate of photorespiration in C3 plants increases significantly with lower $CO_2$ concentrations, but C4 plants are unaffected. Temperature also plays a vital role in the distribution of the plant. C4 plants are called "warm-season grasses" because a higher temperature favors the growth of plants. The light intensity has direct influence and higher light intensities favouring C4 growth.

The light interception (LI) by a canopy is calculated by the difference between the incident radiation and reflected radiation by the soil surface. It is important factor in determining the crop growth and development which provides the energy needed for fundamental physiological processes for example, photosynthesis and transpiration. Plants intercept both direct and diffuse sunlight. The upper leaves get these two kinds of radiation, while the lower leaves intercept a small proportion of direct radiation. Therefore, diffuse radiation turns out to be more significant in the lower leaves because of radiation transmitted and reflected from the leaves and the soil surface. The quantity of radiation intercepted by plant canopy is influenced by a numerous factors such as leaf angle, the leaf surface brightness, the thickness and chlorophyll concentration, which affect the light transmission, the size and shape of the leaf phyllotaxis and vertical stratification, and the distribution of direct and diffuse sun light. Out of 100% energy received by leaf only 5% is converted into carbohydrates for biomass production later.

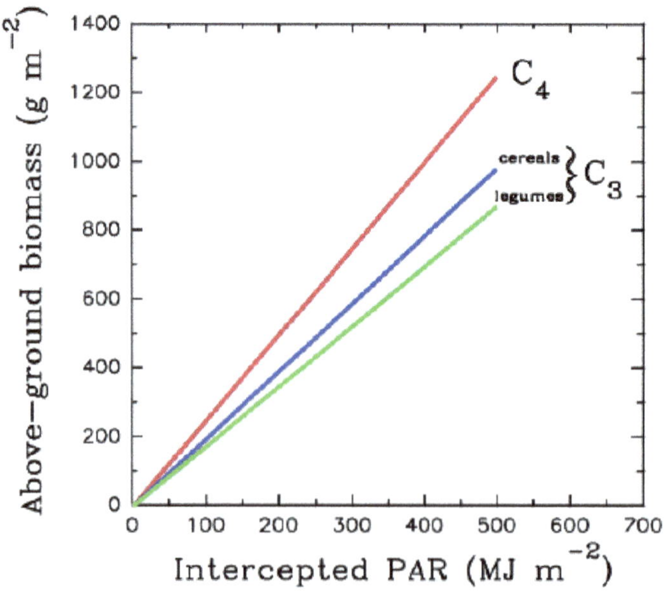


**Figure 33. Typical theorized relationships between cumulated aboveground biomass and cumulated intercepted solar radiation for $C_4$ and $C_3$ species. (Sources: Gosse *et al.*, 1986)**

## Light Interception and Resource Capture in Agroforestry Systems

In agroforestry, tree and agricultural crops are combined together and compete with each other for growth resources such as light, water and nutrients. The resource sharing by the components may result in complementary or competitive effects depending upon the nature of the species, the manner in which they are grown and the climatic factors. Plants and trees may influence its neighbouring species by altering the microclimatic conditions such as temperature, light, wind movement or moisture and by alteration of the balance between beneficial and harmful organism. The important factor to be considered here is the optimum light-use efficiency. The biomass yield limit is set by the available amount of light, its interception and the efficiency with which intercepted light is converted into biomass. Therefore, selection of species in agroforestry must be based on cultural and economic as well as environmental factors. The photosynthesis rate of tree in agroforestry depends on various factors. They are following:

1. **Light availability**- If shading is significant, the understory preference should be for C3 plants as they have a greater efficiency of $CO_2$ uptake at lower radiation than C4 plants. If, however, the overstorey is open, C4 types could be used as understory species (Tieszen, 1983). Photosynthetic capability of different species will undoubtedly be an important physiological consideration in the search for new species and screening of local species for their potential to be used in agroforestry.

2. **$CO_2$ concentration**- The $CO_2$ concentration in the atmosphere affects photosynthetic rates. When the $CO_2$ concentration increases, the photosynthetic rate is also expected to increase. However, the major environmental concern that presently prevails with the adverse effect of an increasing atmospheric $CO_2$ concentration is the possible increase in global temperature (through absorption of infrared of light) and its influence on global weather patterns. Changing climates promise to have a great effect on plant productivity.

3. **Temperature and availability of resources**- The other major factors that affect photosynthetic rates are temperature and the availability of resources like moisture and nutrients. Although agroforestry may cause alteration in the availability of these growth factors. However, plants react differently in their response to the interacting effects of shade and nutrients, and possibly of shade and temperature. Screening crop varieties for their specific responses, and understanding the mechanisms of the responses and manipulating them through management practices will be challenging areas for future research in agroforestry.

4. **Tree management**- The manipulating photosynthesis of plant communities in agroforestry systems through tree management is a major concern. The selection of species along with their arrangement and management determine the photosynthetic efficiency of the whole plant community. Tree management operations like spacing, planting technique, post-planting care, soil moisture and fertility maintenance, canopy manipulation like thinning and pruning, and maintaining healthy and hygiene of plantation influence the productivity of system. The post-planting care of plantation is crucial, depending on the site condition and local biota of pests and competing vegetation. Soil moisture and fertility are two essential parameters of site quality, and they may be effectively modified by irrigation and fertilizer application

to augment nutrients and moisture in short supply.

## Leaf Area Index (LAI)

Leaf is the main photosynthetic functional unit. The leaf area index (LAI) is the concept to estimate the crop's ability to capture the light energy. LAI is regarded as a tool of ecological field and modeling studies. LAI is characterized as the measure of leaf area ($m^2$) in a canopy per unit ground area ($m^2$). It is basic descriptor of vegetation condition in variety of physiological, climatological and biogeochemical studies. For a productive utilization of solar radiation, the major portion of radiation must be absorbed by the photosynthetic tissues of crop. The efficiency of leaf with respect to capture and use of solar energy determines the productivity. The area and arrangement of leaves (the canopy architecture), determine the interception of solar radiation by a crop and the distribution of irradiance among individual leaves (Loomis and Connor, 2002). Leaf area and arrangement change during the life of a crop and/or even during a solitary day by leaf movement influence the interception of solar radiation and which in turn photosynthesis. The maximum crop production needs complete capture of incident light and can only be accomplished with supporting levels of water and nutrients (Loomis and Connor, 2002). The angle, number, size, arrangement, phyllotaxy of leaves decide the photosynthetic area and capacity of individual plants. Multispecies-multilayered plant communities, e.g., homegardens, obviously have multiple strata of leaf canopies and consequently, a substantially higher LAI than in mono-specific stands, which means higher photosynthetic rates. However, higher LAI need not necessarily prompt to higher photosynthetic rates proportionately.

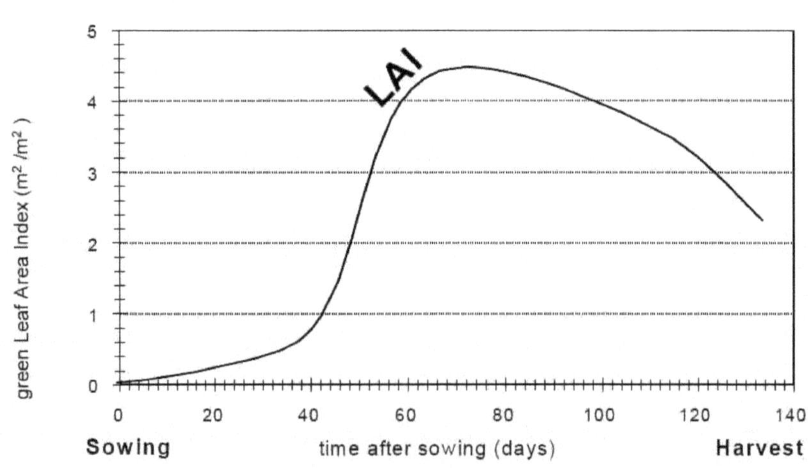


**Figure 34. Typical presentation of the variation in the active (green) Leaf Area Index (LAI) over the growing season for a maize crop (Allen et al., 1998)**

## Methodologies for Biomass estimation in Forestry/Agroforestry Systems

Biomass is produced by living plants converting sunlight into plant material through photosynthesis. Plant converts less than 1% of the available sunlight to stored, chemical energy through photosynthesis. Generally, dry weight of living organism is considered as Biomass. Estimation of biomass is essential for determining the status and flux for biological materials in an ecosystem and for understanding the dynamics of the ecosystem (Anderson, 1970). Swank and Schreuder (1974) have explained that quantity of tree biomass per unit area of land constitutes the primary data needed to understand the flow of materials and water through forest ecosystems. However, the biomass and productivity estimates of tree species vary from place to place due to variation in climate, soil nutrient status, temperature and rainfall (Lodhiyal *et. al.*, 2002).

Biomass study is essential to know the carbon cycle and understanding the dynamics of the ecosystem. Biomass is also regarded as an important indicator of ecological and management processes in the vegetation. Biomass is a reflection of controlling the nutrient, water, and light availabilty on the site. Standing biomass of crop indicates the amount of energy stored in the vegetation, which can reflects the potential productivity at the site. Biomass and residual biomass also strongly influence dynamics of the ecosystem and control the properties of the site including infiltration, runoff, and erosion. Estimations of tree biomass may be required for general forest planning or for studies of energy and nutrients flows in ecosystems. However, currently the emphasis is on tree biomass both as carbon stocks as a means to reduce $CO_2$ concentration and as an energy source. The total biomass of a given stand or forest is generally obtained by multiplying biomass $ha^{-1}$ by the area under consideration. The fundamental prerequisite in biomass estimation is the availability of appropriate species-specific biomass equations for individual trees. In general, a biomass equation may be utilized to estimate the dry weight of biomass for a whole tree or parts of a tree by using tree-level data (tree species, diameter, height, etc.). If only a sample trees within a given unit is measured, there are mainly three ways to estimate biomass $ha^{-1}$. They are:

1. The area unit can be characterized through stand-level variables (e.g. species distribution, basal areal $ha^{-1}$, mean height) and the biomass estimated by using stand biomass equations.
2. Individual tree-level information can be generated by using diameter distribution models and/or height curves, and the biomass can be estimated from equations for individual trees. However, this method requires available information on number of trees $ha^{-1}$.
3. The application of stand-level volume equations and corresponding biomass expansion factors (BEF) to estimate total biomass $ha^{-1}$. BEF is a factor that converts timber volumes into the dry weight of the biomass.

Estimation of biomass in agroforestry is important to know carbon stocks. There are two methods to estimate the tree biomass:

1. Destructive sampling method

2. Non-destructive sampling method

Destructive method is also known as direct estimation of biomass which involves the felling of trees to determine biomass. Tree biomass estimation by destructive sampling is an old approach. It consists of cutting down sample trees, separating into their various components (stem, leaves, braches, etc.), digging out and washing the roots (tap roots and fine roots), determining their samples dry weights of each part, and then adding them up to get the total biomass of tree. Afterwards, allometric equations are developed as regression models using data of diameter at breast height (DBH), total tree height or commercial bole height as the independent variables and total dry weight as the dependent variable. The destructive method of determining tree biomass, though comparatively accurate, is extremely time consuming, high cost and labour-intensive. Indirect methods usually used to study the standing biomass mainly. In this, the estimation of stand biomass is done based on allometric equations using the measurable parameters. The Diameter at breast height (DBH) and height are two most common predictors for aboveground biomass estimation but many studies showed DBH as one of the universally used predictor because it shows a high correlation with all tree biomass components and also measured easily with high accuracy. A linear regression equation for different component of tree was developed and subjected to regression in form of $Y = a+bX$,

Where, Y is dry weight of components (kg) and X is the mean DBH of tree.

Equations for the following tree components were included: total aboveground biomass, stem wood, stem bark, total stem (wood and bark), foliage, and branches (wood and bark). Equations are presented with the range of DBH values of the sample, sample size, fitting method used to estimate the parameters a and b, correction factor for a bias introduced by logarithmic transformation of the data, site index and geographic location of the sampled stand(s), and a reference to the paper in which the equation was published. This is a unique source of equations that can be used to estimate tree biomass and/or to study the variation of biomass components for a tree under different agroforestry system.

Biomass accumulation ratio i.e. BAR (biomass/net productivity) has been used to characterize the production conditions in communities. It expresses the quantity of biomass retained per unit of net production. Das and Chaturvedi (2005) recorded the biomass accumulation ratio (BAR) of 2.1-3.2 for *P. deltoids* plantations which was lower than 4.9-7.7 reported for *P. deltoids* (D-121) plantation in India (Lodhiyal et al., 1995). The low ratio indicated that the system supports a lower total biomass for a given amount of primary production. In the tree layer, the aboveground photosynthetic to non-photosynthetic tissue ratios were 0.05 (young) and 0.06 (mature). Low root: shoot biomass ratio suggested that that the species could be accumulating more aboveground biomass while building the canopy and that they are still in their developmental phase. The root: shoot ratio remains more or less constant and forms the characteristic feature of a species (Penka, 1965).

## Research Challenges in Biomass Estimation

**(1) Biomass estimation for carbon stock versus bioenergy**- The biomass estimation over the past decade has focused on biomass as carbon stocks, while less attention was paid to biomass as an energy source. It is, however, important to be aware of differences between the approaches. In carbon estimations the aim is to calculate the total biomass, but quantification of different biomass fractions is less important. However, estimations related to bioenergy, it is important to quantify the biomass that actually can be used for energy purposes. There is need to distinguish between different fractions since their potential for producing energy differs (Palander *et al.*, 2009)

**(2) Systematic evaluation of existing biomass equations for application in India-** A very few biomass equations are based on data covering a wide range of geographical locations, conditions and treatments. Therefore, research should focus on a systematic evaluation of existing equations. The evaluation should be done in two different ways, *i.e.* based on qualitatively and empirical data. A qualitative approach should include questions related to the representativeness and structure of the data, the statistical methods used, the capability of the equations to describe the different components of a tree. The comparisons could be done for different conditions (e.g. productivity, treatments) and geographical areas.

## Methodologies for Carbon Sequestration Potential of Agroforestry Systems

The process of removing carbon from the atmosphere and depositing it in a reservoir is known as carbon sequestration. C sequestration refers to the rate process involving the time factor (e.g., Mg C $ha^{-1}$ $year^{-1}$), whereas C stock (Mg $ha^{-1}$) does not consider the time factor. Tree growth acts as an essential method for capturing and storing atmospheric carbon in vegetation, soil and biomass. The United Nations Framework Convention on Climate Change has perceived the significance of plantation forestry as a greenhouse gas mitigation option, as well as the need to monitor, safeguard and increase terrestrial carbon stocks (Updegraff et al., 2004). Trees play a vital role in alleviating the diverse effects of climate change, environmental degradation, increasing $CO_2$ in the atmosphere and its consequences. Trees enhance the sequestration of carbon into soil and plant biomass. Consequently, tree based land use practices could be suitable choices to store atmospheric $CO_2$ due to their cost effectiveness, high capability of carbon uptake and associated ecological as well as social benefits (Singh and Lodhiyal, 2009). The projections of the International Centre for Research in Agroforestry suggest that significant funds could potentially be available to finance sustainable rural development and adaptation to climate change, as the carbon market may exceed US $1 trillion by 2025 (ICRAF, 2009).

The role of land use systems in mitigating the rising atmospheric $CO_2$ levels through C sequestration in ABG and BGB became an important area of research during the past decade. Agroforestry attracted special attention due to perceived advantages of the large volume of aboveground biomass (AGB) and belowground biomass (BGB) by deep root systems of trees. The woody perennial-based land use systems have relatively high capacities for capturing and storing atmospheric $CO_2$ in vegetation, soils, and biomass products. Therefore, it is essential to estimate and quantify the biomass production and carbon sequestration in agroforestry systems. While the methods used to estimate carbon sequestration in agroforestry systems vary widely and consequently, there is enormous inconsistency in the available information datasets (Nair, 2011). Moreover, the estimations involve several assumptions, some of which are erroneous (Nair, 2011). Therefore, the methods and procedures adopted in biomass estimation and C-sequestration potential have to be consistent and standardized, so that development plans for the future are based on common standard value. In agroforestry, C sequestration primarily involves the uptake of atmospheric $CO_2$ during photosynthesis and fix C into aboveground (stem, leaves and branches) and belowground (detritus, root, soil organisms) for long term storage. The C sequestration in each compartment varies which is dependent on different factors such as climatic, edaphic, topographic and previous land use pattern, nature and age of tree components). Agroforestry have a higher potential for C-sequestration than pastures or agriculture crops growing under similar eco-region (Kirby and Potvin 2007). It is estimated that on an average, approximately one-thirds in AG and two-thirds in BG (including roots and living biomass) of total C is stored in tree-based land use systems (Lal 2010).

## Measurement of Carbon Sequestration in Agroforestry Systems

Carbon Sequestration involves the process of removing, transfer of atmospheric $CO_2$ and its secure storage in long-lived pools. Agroforestry has a potentially important role to play in climate change mitigation through increased carbon storage in the above ground biomass and below ground biomass. Generally, aboveground biomass is estimated from allometric equations and belowground biomass is considered as a fraction, usually 30%, of AGB, and 50% of the total biomass is taken as C stock (and sequestered C). While, in case of soil, the C content as determined by soil analysis and expressed as C stock. The estimations of C stock in agroforestry as described may have deficiencies and inadequacies arising from both the assumptions used and the procedures adopted. Some of the commonly used assumptions and the errors involved while computation of C-sequestered in AFS are listed below:

- Carbon stock Vs C sequestration: C sequestration is a rate process involving the time factor (e.g., Mg C $ha^{-1}$ $year^{-1}$), but C stock (Mg $ha^{-1}$) doesn't consider the time factor.
- Carbon content related to biomass estimation with an assumption that 50% of the tree biomass is made up by C. While, it is often less than that.
- All biomass represents as C-stock. While, the fact is that all biomass does not end up in "long-lived" carbon pools. The leaf litter decomposes rapidly and releases $CO_2$ back to the atmosphere. The biomass is variable depending on a number of factors including the species, plant part, and edaphic and climatic conditions.

- All C in soil represents C-stock. While, recent additions of litterfall in surface soil are subject to rapid decomposition and release of $CO_2$ with only a small fraction of it getting transformed to stable C in "long-lived" pools. Therefore, there are some complexities to quantifying C sequestration and how it relates to C stocks.
- The management practices on tree growth do not consider while calculating the biomass and C-sequestration. Therefore, the differences in growth form of trees and management practices may lead to under- or over-estimations of root biomass.
- Usually, Amount of C sequestered is taken uniform for a given agroforestry practice. While, high levels of spatial and temporal heterogeneity exist in similar agroforestry practices at different locations. Such extrapolation between similar agroforestry practices can be misleading.

## Operational Inadequacies and Inaccuracies

The most common inconsistency in C stock and C sequestration data in Agroforestry from different locations is related to soil. Soil C stock is conventionally derived by multiplying the analytical data, which is usually in mass per unit mass of soil (g C 100 g soil$^{-1}$) with the soil's bulk density (BD), which is expressed in mass per volume of soil (g cm$^{-3}$ or Mg m$^{-3}$), and with soil depth. The carbon stock in the soil was computed by using the formula,

Soil C stock (Mg ha$^{-1}$) = mass per unit mass of soil (g C 100 g soil$^{-1}$) × bulk density × soil depth (For example, Soil C stock: 2.0 g 100 g$^{-1}$ × 1.0 Mg m$^{-3}$ × 1 m = 200 Mg ha$^{-1}$).

There is an irregularity in this conversion because of BD value involves a volume measure while the C stock value is expressed in an area measure (ha). To overcome this, C stock reported in Mg ha$^{-1}$ is assumed to be for 1 cm thickness (depth) of the soil unless the depth is otherwise specified. Thus, depth should be mentioned while reporting the C stock in agroforestry. Unfortunately, soil C stocks in AFS, either do not mention such details, or do not follow any uniform norm about the depth (Nair *et al.*, 2010). This may lead to confusion and speculation when the data are compiled or compared. Therefore, the impact of soil bulk density on C stocks is particularly important when comparing land use treatments that result in different BD values, as may be the case with AFS compared with annual crops or pastures.

**Sampling depth:** Usually, soil C studies are limited to the surface soils of 20-30 cm depth and it cannot be compared with those of 0–50 cm study. Therefore, sampling beyond the surface soil should be emphasized while studying agroforestry, not only because tree roots extend to deeper soil horizons, but also due to role of subsoil in long-term stabilization of C.

**Sample preparation (sieve size):** The 2 mm sieve is used for preparing soils for laboratory analyses universally. The fractions more than 2 mm in size are often discarded but they might constitute a sizeable amount of the soil and may contain some C (Howlett et al. 2011).

## Carbon Sequestration Potential of Agroforestry Practices in India

In India, average carbon sequestration potential in agroforestry has been estimated to be 25 t/ha over 96 mha area (Sathaye and Ravindranath, 1998), however, these estimates are mostly derived from biomass productivity and often do not include the soil stocks. Soils, however, play a central role in the global carbon cycle and constitute a large carbon reservoir (Lorenz et al., 2006). The soils have been suggested as a largest potential sink for atmospheric carbon which is 3.3-fold larger than the atmospheric-carbon pool and 4.5-fold the biotic pool (Lal, 2004a). Therefore, the woody biomass in the agroforestry plantations and soil stores carbon in substantial amount than conventional crop cultivation. Evidences are now emerging that agroforestry systems are promising to increase above ground and soil carbon stocks to mitigate green house gas emission (Jose, 2009). However, carbon storage potential depended on tree density, age, structure and composition, carbon concentration in plant parts (Kanime *et al.*, 2013).

**Table.7: Carbon stocks in different tree-based agroforestry systems (Mg ha$^{-1}$)**

| Region/Agroforestry system | Species components | Carbon stock (t ha-1 ) | | References |
|---|---|---|---|---|
| | | Components | Total | |
| **Sub tropical region (Central Himalayan)** | | | | |
| Agri-silivicuture (Block plantation) | *Populus deltoides* | 22.81 (AG)<br>5.85 (BG) | 28.67 | Kanime et al., 2013 |
| Agri-silivicuture (Boundary plantation) | *Eucalyptus tereticornis* | 9.55 (AG)<br>0.97 (BG) | 10.52 | Do |
| Agri-silivicuture | *Dalbergia sissoo* | 35.77 (AG)<br>7.62 (BG) | 43.39 | Do |
| **Temperate region (Himachal Pradesh)** | | | | |
| Horti-pastoral | *Mangifera indica/Prunus salicina* + grasses | 17.43 (AG)<br>5.04 (BG) | 22.47 | Minj, 2008 |
| Silvi-pastoral | *Leucaena leucocephala / Populus deltoids* + grasses | 22.63 (AG)<br>7.18 (BG) | 29.82 | Minj, 2008 |
| Agri-silvi-culture | Wheat-paddy + *P. deltoids / Bauhinia variegata/* Grewia optiva | 30.54 (AG)<br>9.43 (BG) | 39.97 | Minj, 2008 |
| **Semi-arid region (Jhansi)** | | | | |
| Silvipasture | *Acacia nilotica/ Dalbergia Sisoo/ Hardwickia binata* + Pasture | 9.5-17.0 (AG) | - | Rai et al., 2001 |
| Agri-silviculture | *Dalbergia sissoo* + Crop (11 year old) | 26.0 | | NRCAF (2007) |
| **Arid region (Rajasthan)** | | | | |
| Horti-silviculture | Z. mauritiana + P. cineraria | 1.25 (AG)<br>2.25 (BG) | 3.50 | Singh and singh, 2015 |

| Region/Agroforestry system | Species components | Carbon stock (t ha-1 ) | | References |
|---|---|---|---|---|
| | | Components | Total | |
| Shelterbelt AGF (Karnataka) | *Pongamia pinnata, D. sissoo, Acacia auriculiformis, Tectona grandis, Casuarina equisitifolia, Azadirachta indica* | 11.34 (AG) 2.54 (BG) | 13.88 | Swamy et al., 2017 |
| **Tropical region (Homegarden)** | | | | |
| Mizoram (N-E region) | Mixture of different species | - | 108.83-258.43 | Singh and Sahoo, 2015 |
| Kerala | Mixture of different species | 16-36 | - | Kumar,2011 |

## REDD + Actions and its Relevance to Indian Policy

Reducing emissions from deforestation and forest degradation, alongwith conservation and sustainable management of forests, known as REDD+ is recognized as an innovative way to mitigate climate change under UNFCC framework and also emphasize on afforestation and reforestation to sequester more carbon dioxide from atmosphere. Agro-forestry can serve as best option for REDD plus agro-forestry has been reported to sequester more carbon dioxide than other agricultural systems like. Agroforestry practices can sequester more carbon as they utilize natural resources more efficiently and perennial fast growing trees capture more carbon in biomass, for long term and at faster rate (Nair et al., 2010). To execute specific REDD+ activities in the nation, *viz.* (a) Reducing emissions from deforestation; (b) Reducing emissions from forest degradation; (c) Conservation of forest carbon stocks; (d) Sustainable management of forest; and (e) Enhancement of forest carbon stocks; some key interventions are needed to be taken for forest management in the country.

## Agroforestry for Creation of Carbon Sinks Outside Forests

Most of the timber and fuelwood necessities in country are being met from trees outside forests, particularly from agro-forestry. These areas could also be taken beneath the "Enhancement of Forest Carbon Stocks" component of REDD+ and can support a program of supply raising quality seedlings to fulfill the requirement of farmers, as well as transportation to the villages to supply straightforward, easy, energy efficient reach. It might additionally support for certification of seed and genetically improved clonal planting material to raising carbon stocks outside traditional forest areas. e.g. trees planting on non agricultural rural lands like homesteads, college yards, compounds of assorted offices, and private/public institutions, public areas, road side, canal side, along the railway lines etc. have tremendous potential for REDD+ mechanism. These areas have the potential of increasing the carbon stocks by 2.41 million tons Carbon and 8.14 m tons of $CO_2$-assimilation (MoEF, 2010). The promotion of energy security and carbon emission reduction objectives should be carried out through energy proficient gadgets and alternative energy sources like bio gas. The backward and disadvantage area burdened with unsustainable harvesting and use of fuel wood should be primarily targeted in order to reduce pressure on the forests. The use of improved cooking stoves, smokeless and energy efficient chullah should be promoted, which could save fuel wood. This would lead to saving of 2 million tons of fuel wood every year (MoEF 2010).

## Synergies with other :Programmes and Arranging for REDD+ Finance

REDD+ can be integrated with other government programmes at national and state levels. Potential outcomes of creating specialized, monetary and social collaborations with MNREGA, Green India Mission and other government schemes should be explored for effective usage of REDD+ in the country. At policy level, technical and financial collaborations should be created between the implementation of three Rio conventions of UNFCCC, UNCCD and CBD in India for efficient management of forestry resources. Local communities that help in achieving the mitigation must be incentivize under the REDD+ projects. The international funds like Forest Carbon Partnership Facility (FCPF) may be explored to incentivize communities to boost networks in the country for REDD+ activities.

## Scope for Enhancing Biomass Production and C-storage through Agroforestry in India

Enhancing biomass production is need of the hour for meeting demand of raw material of wood based industries, bio-energy, fuel wood, fodder and for enhancing carbon sequestration and c-stock. In India, there is huge scope of agroforestry practices in view of enhancing biomass production and this can be achieved by extending area under agroforestry and improving existing and traditional agroforestry practices. There is lot of scope for extending area by including wasteland and marginal land under agro forestry practices, and this will not only meet demand for wood based raw material fodder, fuel wood NTFPs etc. but will also help in achieving the target of covering 33% land area of India under forest and tree cover. The rising pollution and contamination of water, soil with heavy metals can also be tackled by agroforestry sustainably along with high biomass production.

For extending area under agroforestry for higher biomass production and carbon sequestration, the panchayati land, farm bunds, tree less patches around roads, ponds and railway tracks, community land can be targeted to raise agroforestry plantations. Another huge scope is by reclaiming and improving production capacity of wasteland which is 30 to 175 Mha due to environmental degradation because of soil erosion by water and wind, soil salinization /alkalization /acidification, water logging, mine affected areas etc. Improvement of existing and traditional agroforestry practices focuses on scientific planning and establishment of agroforestry systems which includes:

- Choosing suitable tree species (Technically feasible, Ecologically adaptable and economically viable)
- Raising plantation using plus tree or improved genotypes of higher biomass production
- Using quality planting material of trees.
- Using suitable management practices to reduce tree-crop competition for

efficient resource utilization.

- Short rotation forestry (SRF) that attain exploitable diameter within 5-10 years and mean annual increment of at least 15 $m^3$ per ha per annum or 60 cm height per annum can be deployed for higher biomass production. Incorporating SRF will be a win-win situation for higher biomass production along with early and higher economic returns to the farmers.

## Way Forward

- Convergence of all the schemes, policies related to agroforestry to one platform to systematic working and easy mobilization of funds.
- Ecological zone wise prioritization of economically and commercially important tree species by involving research institutes and experts.
- Utilization of barren land, wasteland, canal sides, road sides, railway llines and village community land for establishment of plantation.
- Focusing on tree breeding for developing genotypes with higher biomass production potential.
- Involving farmers by creating awareness for climate change and benefits of adopting agro-forestry practices.
- Linking of wood based industries with farmers for easy sale of products and revenue generation to farmers.
- Evaluating and monitoring of existing agroforestry system in India for improvement and to assess their biomass production and carbon storage potential.

# Chapter 11

# Agroforestry for Watershed Management

*Deepshikha Nirala, Dhanush C. Reddy, Poonam, Aman Dabral and Prashant G Naik*

## Introduction

Agroforestry in India contributes to the target set by the Indian Council of Agricultural Research for increasing forest cover from the present level of 23% of land area to 33%. The Report of the Task Force of Greening India for Livelihood Security and Sustainable Development has suggested that 10 million ha of irrigated land and 18 million ha of rain-fed land should be managed under agroforestry systems. Agroforestry as a traditional land-use adaptation may potentially support livelihood improvement through the simultaneous production of food, fodder and firewood. Environmental benefits include food availability, security of land tenure, increased farm income, restoration and maintenance of biodiversity, and maintenance of watershed hydrology and soil conservation. Agroecosystems can be designed to assist the adaptation of communities and households to in the local and global change. Agroforestry has an important role in reducing vulnerability, increasing resilience of farming systems and buffering households against climate related risks. It can also maintain or increase the supply of ecosystem services – water, soil health and biodiversity, especially under a changing climate, that will continue to support social, industrial and ecosystem services needs. A livelihood is a means of deriving a just and dignified living by the society, family and individuals. It comprises of the assets available to households (human, financial, physical, natural and social capital), the activities, and the access to these (mediated by institutions and social relations) that together determine the living gained by the society, households or individuals. Livelihood improvement through natural resource management seeks to understand individual or household strategies through which long-term progress is made towards a better quality of life.

In addition, agroforestry practices have been intertwined with the various programs/schemes like watershed development, rehabilitation of problem soils, treatments of degraded and other wastelands etc. The area under forest is degrading due to demographic pressure and infrastructure growth needs while the agricultural area is almost stable. In India, nearly 120.72 million ha land or 37% of the total geographical area is under one or other forms of soil degradation (e.g. water erosion: 93 million ha (M ha), wind erosion: 11 M ha, salt affected soils: 6.74 M ha and 16.53 M ha of open forest area). Up to March 2007, 56.54 M ha has been treated under various watershed development programs; however, a sizeable area is yet to be treated.

## Watershed Management

Watershed management is the process of guiding and organizing the use of land and other resources in a watershed to provide desired goods and services without adversely affecting soil and water resources. Each project under the program is a micro – level effort to achieve this objective by treating the under productive or unproductive land and taking up allied activities for the benefit of the landless. The program adopt a common strategy of multi resource management involving all stakeholders within the watershed who, together as a group, co-operatively identify the resource issues and concerns of the watershed plan with solutions that are environmentally, socially and economically sustainable. The three main aims of the watershed programs are: (i) Enhancing crop production and increasing farmer's income (ii) Minimizing degradation of the natural resource base and (iii) Improve livelihoods of the rural people in the watersheds.

## Watershed

A watershed, also called as a drainage basin or catchment area, is defined as an area in which all water flowing into it goes to a common outlet. People and livestock are the integral part of the watershed and their activities affect the productive status of watersheds and vice versa. From the hydrological point of view, the different phases of the hydrological cycle in a watershed are dependent on the various natural features and human activities. Watershed is not simply the hydrological unit, but also socio-political-ecological entity which plays a crucial role in determining food, social, and economical security and provides life support services to rural people.

- Watershed management is increasingly being recognized as the ideal approach for integrated natural resources management in rain fed areas. About 51% of India's geographical area (329 million ha) are categorized as degraded, most of which occur in rain fed agro-eco systems.
- About 70% of the population is dependent on agriculture, and two thirds of the cropped area is dependent on rainfall without any protective irrigation. The rain-fed areas are the hotspots of poverty, malnutrition, food insecurity, prone to severe land degradation, water security and poor social and institutional infrastructure.
- Watershed development program is, therefore, considered as an effective tool for addressing many of these problems and recognized as a potential

engine for agricultural growth and development in fragile and marginal rain-fed areas.

- Management of natural resources at watershed scale produces multiple benefits in terms of increasing food production, improving livelihoods, protecting environment, addressing gender and equity issues along with biodiversity concerns.
- Participation of people is essential for the success of watershed program. Participatory watershed management is a process which aims to create a self supporting system which is essential for sustainability.
- The concept of Participatory watershed management emphasizes a multi-disciplinary and multi-institutional approach. The process begins with the management of soil and water, which eventually leads to the development of other resources.
- Human resources development and large-scale participation are essential since finally it is the people who have to manage their resources. Peoples or farmer participation is key to the success of any watershed development program.
- Moreover the watershed should be used as an entry point for convergence of various natural resource management-based activities for improving the livelihoods rather than soil and water conservation programs.

## Principles of Watershed Management

There are some basic principles of watershed management are as follows:

1. Utilizing the land according to its capability.
2. Adequate vegetative cover during the rainy season.
3. Conserving as much rain water as possible where it falls.
4. Draining out excess water and diverting it to storage ponds.
5. Avoiding gully formation and checking at suitable intervals to control soil erosion and recharge ground water.
6. Maximizing productivity per unit area, per unit time and per unit of water.
7. Increase cropping intensity and land equivalent ratio through intercropping and sequence cropping.
8. Safe utilization of marginal lands through alternative land use systems.
9. Ensuring sustainability of the ecosystem benefiting the man-animal-plant-land-water complex in the watershed.
10. Maximizing the combined income from the interrelated and dynamic crop-livestock-tree labor complex over the years.
11. Stabilizing total income and cutting down the risks during aberrant weather situations.

12. Improving infrastructural facilities with regard to storage, transportation and marketing.

Some of the Major Objectives of Watershed Management are to:

- Conserve soil, rain water and vegetation effectively and harvest the surplus water to create water sources in addition to ground water recharge.
- Promote sustainable agriculture and stabilizing crop yields by adopting suitable soil, water, nutrient management and crop management practices.
- Cover the non-arable area effectively through afforestation, horticulture and pasture land development based on land capability class.
- Enhance the income of individuals by adopting alternative enterprises.
- Restore ecological balance.

## Watershed Approach to Conservation and Development

A watershed is physically distinguished into lowland and upland or, in conventional terms, upstream and downstream. These two physical entities are integrated through a number of bio-physical and socio-economic linkages. Rivers originate in the uplands and provide water for drinking purposes as well as for the pursuit of economic activities in the entire watershed. Thus, any change in the volume of water caused by human activities leads to chain reactions in both upstream and downstream areas. Increased surface runoff caused by upland degradation raises the volume of water in rivers and streams, and accelerates the removal of soil that is deposited in reservoirs, riverbeds and canals. As a consequence, the upstream area major issues include dwindling land productivity and the downstream area major issues include flooding. Some agricultural lands, however, in lowlands benefit from flooding, as through this process nutrients are deposited in the rich silt. Siltation of particularly reservoirs constructed for hydroelectricity generation and irrigation impairs the amounts of energy and irrigation water provided by these structures and eventually curtails both agricultural and nonagricultural economic activities mainly in the downstream region. Agricultural activities in the upstream area may not be much affected by such change, as most of these activities take place on hill slopes with rain-fed cultivation system. Reduced volumes of water in rivers caused by the development of irrigation schemes in the upstream area also generate a chain of negative effects particularly in the downstream.

The specific advantages of a watershed approach to conservation and development are the following:

i. Conservation and development activities in upland and lowland can be planned and implemented in a well-coordinated way, so that the entire watershed benefits from the activities.

ii. The above facilitates cost-effective implementation of conservation and development activities.

Both upland and lowland settlers are the primary beneficiaries of watershed resources. Thus, in principle, both groups should have an interest in watershed management, provided costs and benefits are well understood and the planning and management system is participatory.

## Sustainable Rural Livelihoods in Watershed

- A livelihood comprises the capabilities, assets (including both material and social resources) and activities required for a means of living. A livelihood is sustainable when it can cope with and recover from stresses and shocks and maintain or enhance its capabilities and assets both now and in the future, while not undermining the natural resource base."
- Sustainable livelihoods approach identifies five assets - physical, natural, social, financial and human capital - that people draw upon in order to construct livelihood strategies. An underlying principle of the livelihoods approaches is that they begin from a position of assessing people's strengths rather than needs. Thus, in order to earn sustainable livelihoods, the poor have access to five different types of capital:

**Natural Capital:** It refers to the availability of water, forests, land, climate and other natural resources.

**Physical capital:** This is the basic infrastructure, goods and services needed to support livelihoods (e.g. water supply and sanitation systems, affordable transport, shelter, energy, etc.)

**Social capital:** It refers to relationships of trust and reciprocity that support cooperative action, membership of formal and informal groups and networks that increase people's ability to work together and access institutions and services. Formal law (statutory and religious) and informal law (customary) can also be seen as forms of social capital.

**Human capital:** refers to the skills, knowledge, beliefs, attitudes, ability to labor and good health that enable people to pursue different livelihood strategies.

**Financial capital:** refers to the financial resources including savings, credit provision and regular inflows of money (e.g. wages, remittances, subsidies, etc).

## Watershed Development Programs in India

Water shed development originally managed by the national wasteland development board under the Ministry of Environment and forest. It is now placed under the Ministry of Rural Development and Department of Land Resources. The main objective of this program for development of waste lands in non-forest areas, checking of land degradation, putting such waste land into sustainable use and increasing biomass, availability of fuel wood, fodder and restoration ecology, etc. There are various watershed projects/ programs in India are enlisted in Table 8 from 1962 up to 2006. Thus the concept of watershed development is an integrated nurture with multi-disciplinary activities in the area.

At the present Ministry of Rural Development and Department of Land Resources, Government of India funding watershed development program under D.P.A.P., D.D.P., and Integrated Wasteland Development Plan (I.W.D.P.) etc. This program is intended to be taken up in rain-fed and drought-prone areas especially predominated by SC/ST population and preponderance of wasteland. There are six major projects/ programs in watershed development program namely,

i. National Watershed Development Project for Rainfed Areas (NWDPRA)

ii. Watershed Development in Shifting Cultivation Areas (WDSCA)

iii. Drought Prone Areas Program (DPAP)

iv. Desert Development Program (DDP)

**Table 8. An overview of Watersheds in India:**

| Year of establishment | Watershed Projects/ Programs |
|---|---|
| 1962-63 | River Valley Projects – arrest silt |
| 1970-71 | Rural works prog – drought relief funds |
| 1973-74 | DPAP – Work by govt's soil conservation dept – scattered over larger areas |
| 1974-75 | WDPSCA – shifting cultivation areas |
| 1977-78 | Desert Development Program (DDP) |
| 1980 | ICAR model watersheds |
| 1982 | Swaminathan task force –watershed management as principle |
| 1985-86 | RAS – reclamation of alkali soils |
| 1987 | Central Sanctioning Committee – SWC emphasis |
| 1988-89 | IWDP – integrated watershed development program |
| 1989-90 | NAEP - National Afforestation and Eco - Development project |
| 1990-1991 | NWDPRA National Watershed Development Programme for Rainfed Areas |
| 1994 | Technical committee on DPAP and DDP – integrated and people centred watershed development approach |
| 1999-2000 | WDF - Watershed Development Fund |
| 2006 | Parthasarathy Committee report |

i. Integrated Wasteland Development Project (IWDP)

ii. Employment Assurance Scheme (EAS)

These six projects/ programs also account for about 70 % of the funds and area under watershed programs in the country. Through various watershed development programs, about 30 million ha. Of land has so far been developed at an expenditure of Rs. 9343 crore, including external funding up to the end of IX Five Year Plan. During X Five Year Plan about 11.4 million ha. is proposed to be developed at an outlay of Rs.7440 crore. Besides, an area of 1.24 million ha. is likely to be treated under watershed programs at a cost of Rs.1872 crore through ongoing externally aided projects.

## Improving livelihoods for the rural poor is usually understood as a basket of options to increase earnings by:

- Improved access and control over common property resources like forests and pastureland –so that they can gather grass and non-timber forest produce (NTFP) to feed their animals or sell

- ❑ Improved short term and long term wage labor opportunities
- ❑ Increased economic security and access to services
- ❑ Improved access to the government schemes through better information
- ❑ Appropriate support for marginal farming operations on high or sloped land, access to new technology and resources
- ❑ Creation of productive assets
- ❑ Improved access to the government schemes (including social welfare schemes)

**Reduce costs by:**

1. Improved food security, off-farm employment and reducing seasonal vulnerability
2. Improved access to savings and credit facilities
3. Better ability to withstand disasters (sickness and death in the family, vulnerability to drought, flood, cyclones, etc)
4. Increased local control over water resources and facilities

**Improve quality of living by:**

- ❑ Improved access to water for drinking and domestic uses – so as to improve the quality of life and water availability for crop production.
- ❑ Reduced drudgery (e.g. time taken on fuel, fodder and water collection)
- ❑ Improved livelihoods health status, particularly for those involved in hazardous occupations (e.g. slate quarries)

**Empower rural poor by:**

- ❑ Active involvement of the poor in decision making and planning processes
- ❑ Improved access to public life
- ❑ Improved access to training and skills up-gradation opportunities
- ❑ Improved dignity and reduction of oppression and elimination of discrimination
- ❑ Improved access to government schemes through better information and support
- ❑ Enhanced ability to implement project activities through groups
- ❑ Increased opportunities to learn and acquire additional exposure

**Improved watershed management offers many potential benefits for the local community, and the larger cross section of society**

**For the local Community:**

- Lower land development costs.
- Reduced flooding and water logging.
- Reduced soil erosion and land degradation.
- Increased agricultural productivity.
- Improved livelihood options.
- A more dependable, clean water supply for domestic and industrial use.

**For the larger cross section of society:**

- Less danger from floods to downstream cities and farmlands.
- Reduced sedimentation of costly irrigation projects.
- Better conservation of natural resources.

## Soil and Water Conservations Measures for Watershed management

It is the technique in which deterioration of soil and it loses is reserve by using it within its capabilities and applying conservation technique for production as well as improvement of soil.

**Soil and Water Conservation Measures**

| Agronomical Measure ( Biological) | Engineering Practices |
|---|---|
| 1. Contour cultivation | 1. Terracing |
| 2. Strip Cropping | a. Diversion terrace: |
| a. Contour strip cropping | i. Magnum type. |
| b. Field strip cropping | ii. Nichols type. |
| c. Buffer strip cropping | iii. Broad based type |
| d. Wind strip cropping | iv. Narrow based type |
| 3. Tillage practices | b. Retention Terrace |
| | c. Bench terrace |
| a. Mulch Tillage | |
| b. Vertical mulching | 2. Bunding |
| c. Minimum tillage | a. Contour banding |
| d. Conventional tillage | |
| e. Listening | i. Narrow based |
| | ii. Broad based |
| 4. Soil management practices | |
| 5. Supporting Practices (Interplanting, fertilizer application) | c. Side bunds |
| 6. Vetiver grass planting | d. Lateral bund |
| | e. Supplemental bunds |
| | f. Marginal bund |
| | g. Shoulder bund |

## A. Agronomical measures:

Agronomical measures or practice is growing crops on non-mild slopes so as to cover the land for reducing the soil erosion. Agronomical measures include contouring, strip cropping and tillage practices, etc. are adopted to control the soil erosion. The use of these measures is entirely dependent upon the soil types, land slope and rainfall characteristic. It plays a second line of defense after mechanical or engineering measures. It is more economical, long lasting and effective in controlling the soil erosion.

### 1. Contour cultivation:

It refers to all the tillage practices, mechanical treatments like planting, tillage and intercultural, performed nearly on the contour of the area applied across the land slope. Inflow rainfall regions the primary purpose of contour cultivation is to conserve the rain water in the soil as much as possible.In humid regions its basic purpose is to reduce the soil erosion or soil loss by retarding the overland flow. In this system, the furrows between the ridges made on the contours hold the runoff water and stored them into the soil. Thus they reduce the runoff and soil erosion.

### 2. Strip Cropping:

It is also a kind of agronomical practice, in which ordinary crops are planted or grown in the form of relatively narrow strips across the land slope. These strips are so arranged, that the strip crops should always be separated by strips of close-growing and erosion resistance crops. Strip cropping checks the surface runoff and forces them to infiltrate in the soil, which facilitates to the concentration of rain water. It is more effective than contouring [about twice effective as contouring] but it does not effect on soil erosion. It controls erosion by

- Reducing the runoff flowing through the close growing sod strips.
- Increasing the infiltration rate of soil under cover condition.

*Types of strip cropping:*

a. Contour strip cropping.
b. Field strip cropping.
c. Buffer strip cropping.
d. Wind strip cropping.

### 3. Tillage practices:

It is defined as mechanical manipulation of soil to provide a favorable environment for good germination of seed and crop growth, to control the weeds, to maintain infiltration capacity and soil aeration. Tillage practice protects and maintains a strong soil structure to fight against erosion.

*Types of tillage operation [practices]*

**a. Mulch tillage:** It is the application of plant residues or other material to cover top soil surface.

**Mulching material:** Cut grasses, straw material, wood chips, saw dusts, paper and sand stones, glass wools, metal foils and stone plastic.

**Types of Mulch:** Natural, Synthetic, Petroleum, Conventional, Inorganic, Organic.

**b. Vertical Mulching:** Insertion of stuffed plant residue vertically into sub-soiler marks to keep the slot open.

**c. Minimum Tillage:** Preparation of seedbed with minimum disturbance of soil.

**d. Conventional tillage:** Ploughing, secondary cultivation with harrowing and planting.

**e. Listing:** Used for controlling soil erosion.

i. Hard ground listing.

ii. Loose ground listing.

**4. Soil management practices:**

Various soil and land management practices are –

a. Those practice which helps to maintain the soil filtration rate at high level to reduce runoff to a negligible amount.

b. Practices which helps in safe disposal of runoff from the field.

The cultural practices which are helpful for creation of high infiltration rate are essential based on farming techniques, tillage or minimum tillage and use of cover crops. Where as the safe disposal of runoff from the field is carried out by physical manipulation of soil surface. Including land shaping, leveling construction of ridges, bunds and water ways

**5. Supporting Practices:**

It involves application of fertilizers to soil either to make more fertile or to recover the fertility loss during different physical action. Application of fertilizer plays sometimes a significant role to developed abundance vegetative growth e.g. grass waterways and terrace outlet are generally established on low – fertile sub soil. Inter planting refers to seeding of grass or legume crops in combination of maize or other crops to achieve better result on erosion control.

**6. Vetiver Grass Planting:**

It is most effective vegetative material foe soil and water conservation, land rehabilitation and embankment stabilization. Vegetative hedge formed with thick growth of vetiver grass forms a protective barrier across slope which slows down sheet erosion and deposit the slit behind hedges.

**B. Engineering Practices:** It is used to control the soil erosion in highly sloped areas

**1. Terracing.** In agriculture, a terrace is a piece of sloped plane that has been cut into a series of successively receding flat surfaces or platforms, which resemble steps, for the purposes of more effective farming. This type of landscaping is therefore called terracing. It is of following types;

**a. Diversion terrace:** A diversion is designed to catch and hold all the water flowing into it, which usually includes a substantial amount of foreign drainage. It is of following types:

- ❑ Magnum type.
- ❑ Nichols type
- ❑ Broad based type.
- ❑ Narrow based type.

**b. Bench terrace.** Bench terraces are a series of level or virtually level strips running across the slope at vertical intervals, supported by steep banks or risers.

**2. Bunding:** It is of two types

**a. Contour banding:** It consists of building earthen embankments across the slope of the land, following the contour as closely as possible. A series of such bunds divide the area into strips and act as barriers to t5he flow of water, thus reducing the amount and velocity of the runoff.

- ❑ Narrow based
- ❑ Broad based

**b. Peripheral bunds:** Bunds area also constructed along field boundaries without reference to contour. These bunds are called peripheral bunds. They serve as fences, and give protection from water and wind erosion in low rainfall areas. They are not suitable in heavy rainfall areas.

## Role of Agroforestry in Watershed Management

The system of agroforestry involves growing tree with crops on permanent basis in order to meet food, fuel, fodder and fibre needs of the farmers. It has the following aims.

1. Biomass production: The maximum production of biomass per unit area in time is the primary objective of AF in watershed management.
2. Conservation: The barren portion and wasteland of the watershed can be provided with a cover through agroforestry interventions in the watershed management programme to protect soil from erosion.
3. Soil improvement: Agroforestry practices enrich soil by nitrogen fixation and addition of organic matter.
4. Moderation of microclimate: The micro-climate of an area can be moderated by adopting agroforestry systems and in association with the field crops. It has pronounced effect is observed in semi-arid and arid regions.
5. Agro-based cottage industries: An agroforestry programme in watershed management helps in promotion of agro-based cottage industries, such as paper pulp, herbal drugs, fibre, poultry, piggery, dairy, bee-keeping, sericulture cultivation etc. This will provide gainful employment and raise the standard of living of small and marginal farmers.

6. Flood control and drought mitigation: To help farmers build livelihood resilience to floods and drought, agroforestry needs to provide benefits during these events. There are large numbers of scientifically documented environmental benefits of trees during floods. Indeed, environmental benefits comprised half of the listed benefits of trees during floods. During drought, agroforestry's environmental benefits can be enhanced by its ecological characteristics. For example, some tree species can maintain production during drier years because their deep root systems are able to use a greater soil volume to obtain water and nutrients. This characteristic makes certain tree species more resilient to droughts than are cash crops with shallow roots, thus building the five livelihood capital assets. The major livelihood benefits during both floods and drought were fruit for sale and consumption. Agroforestry can provide an alternative source of food and income during floods and drought. These results support the idea of agroforestry as an alternative livelihood that is important during times of stress. Having agroforestry as a supporting livelihood that can still be mused during floods and drought means that agroforestry can help build livelihood resilience by better preparing households tom manage the effects of shocks, to navigate uncertainty, and to adapt to changing environmental conditions.

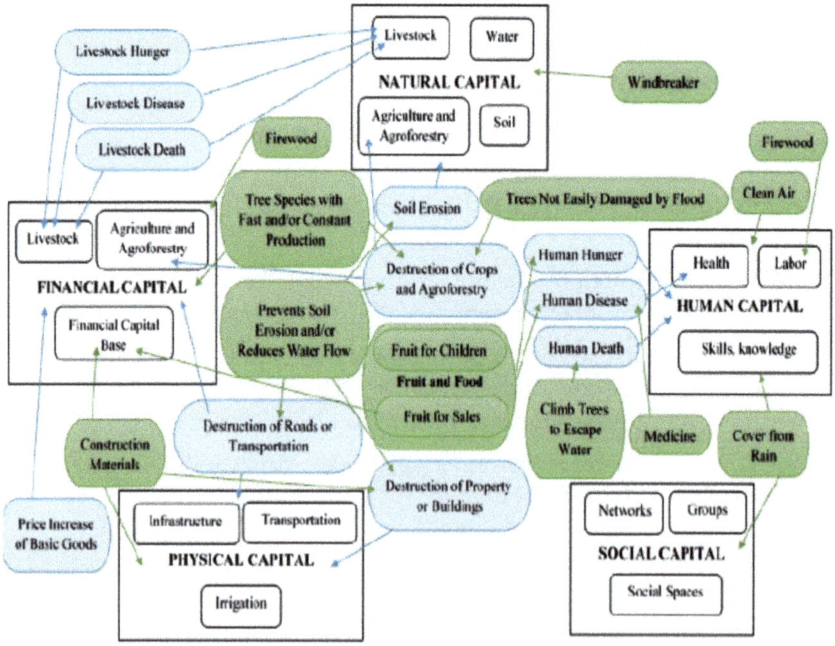


**Figure 35. Flood prevention model in a watershed (Source: Quandt et al., 2017)**

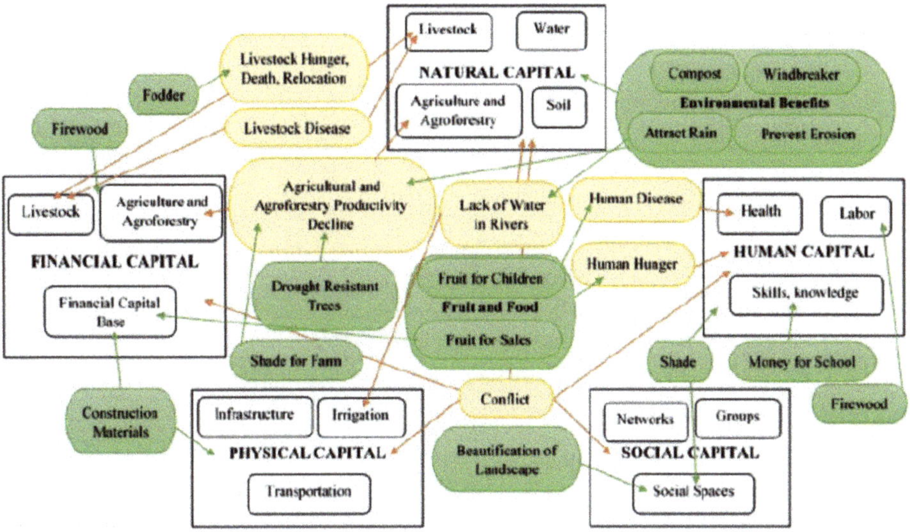


**Figure 36. Drought mitigation model in a watershed (Source: Quandt et al., 2017)**

## Conclusion

Agroforestry systems offer opportunities for the improvement of the livelihood of people through provision of economic and sustained utilization of natural resources. These natural resources can be best utilized and managed in an effective and gainful manner through watershed approach. Watershed management encompasses the concept of optimum utilization of soil and water for agricultural production as well as forestry and applying together the technical know-how, the people and the environment into a harmonious situation. Both natural resources and the socio-economic situations of any watershed are to be given equal importance. Watershed approach is inevitable in developmental programs involving collection, storage and recycling of runoff water. However, suitable eco-friendly methods to the local populace and the situations are to be applied for true and all round agricultural development.

# Chapter 12

# Industrial Agroforestry

***Poonam, Aman Dabral and Sneha Dobhal***

## Introduction

Prior to the enactment of the forest conservation act and national forest policy 1988, most wood industries were deriving their raw material predominantly from forests. But now the forest conservation act and national forest policy 1988 has directed the wood based industries to establish their own plantations through direct relationship with farmers and individuals by providing necessary inputs, technologies, credit facilities and assuring price supportive mechanisms (Durairasu *et. al.*, 2014). Farmers all over India appreciate the multifarious benefits from trees and they have for decades maintained sporadic naturally grown trees on their farms. Mostly land owners prefer to lease their lands on annual basis. Some of the most important considerations which facilitate farmers to take positive decision in favour of Farm- forestry as a preferred land use option are high productivity and far better returns compared to agricultural crops. Tree farming with eucalyptus has now become so popular that irrigated and fertile land of the farmers in Gujarat with JK Paper unit- CPM is producing high productive, disease resistant and site specific clones of *Eucalyptus, Leucaena* and *Casuarina* to the tune of 10 Million per annum and provides technical guidance for planting and maintenance and provides assured buy back of wood at prevailing market price. (Agarwal *et.al.*, 2017). There are also a number of industries based on NTFPs such as beedi (country cigarette), lacquerware, brooms, essential oils, katha and cutch, tannins, resin and rosin, cane and bamboo furniture, herbal medicines and cosmetics, etc. It is estimated that NTFPs worth Rs. 350 billion are used annually in India and the government revenue from NTFPs is around Rs. 20 billion, nearly 50% of the total forest revenue. Total NTFP exports (raw material as well as finished products) were estimated to be US$ 480 million in 1991 (GoI 1999) (Saigal and Bose, 2003). Globally paper Industry is one of the high priority industries having a bearing on the socio-economic development. In India too this industry plays a vital role in the overall industrial growth. Indian paper industry is one of the world's fastest growing industries. It grew at a compounded annual growth rate of 6.7% over FY 06-11. Among the top producers of paper, India ranks at 20th position with an estimated production of 10.9 Mtpa. The Indian sector accounts for about 2.6% of the global production, estimated at 390 Mtpa. In 2012, India recorded paper consumption of 9.3 kg/capita vis-à-vis global average of 58 kg/capita. Currently there are 759 pulp and paper mills with an installed capacity of 12.7 Mtpa, producing around 10.90 Mtpa paper, paper board and newsprint. The production is anticipated to grow up to 14.0 Mtpa by the year 2016 (Report paper and pulp industry, GOI, 2014) and 4.16 million workers were employed in bidi manufacturing in 2005–2006. Of this, 3.42 million workers were engaged in full-time work while 0.74 million workers were employed part time. The bidi sector accounts for about 11% of unorganized manufacturing employment, and full-time workers comprised about 0.74% of total employment in India, estimated at about 457 million formal workers in 2004. At the state level, West Bengal (987 000) and Madhya Pradesh (703 000) employ the most bidi workers, while Madhya Pradesh (40%) and Jharkhand (33%) account for the highest employment contribution as a share of all un organized manufacturing (Nandi, *et al.*, 2014). The use of renewable energy such as solar energy, hydropower, wind etc. can be traced back to the beginning of the modern civilization. But, the concept of bio-fuel or bio-diesel was used in later part of the previous century. Here we consider the biomass, as the source of renewable energy, which is converted to methanol or ethanol etc. and mixed partially with diesel (up to 5%) to use in diesel engines (in automobiles, industries etc.). In future, the bio-fuel may be used without mixing with existing fossil fuel such as diesel (B100) or petrol etc. (Swain, 2014). Yet farmers have amply demonstrated that many of them are keen to opt for agro-forestry plantations if there are major benefits and substantially higher returns from this option (Agarwal *et.al.*, 2017).

**Major Wood Based Industries in India:** Major Wood Based Industries in India are Paper and paper board, News print, Construction Industry, Furniture, Packaging, Rayon grade pulp Automobiles, Agriculture implements, Railway sleeper, sports goods, handicraft, plywood, veneer particle board, MDF board, match box, mining, catchment, pencil industry, etc. The demand of wood for these industries is increasing at alarming rate but there is no concomitant development of plantations which resulted in massive imports. The total industrial wood demand for the year 2020 has been projected to 153 million $m^3$ which will increase from 58 million $m^3$ in the year 2000 (Durairasu *et. al.*, 2014) (Table 9). The demand for wood has been categorized into short rotation species and long rotation species. The short rotation species are primarily used for pulp and paper industries, packaging, sports goods plywood, match and other miscellaneous industries. In general most of the short rotation species are softwoods used in wood industry (Durairasu *et. al.*, 2014). In 2015, growth in the production volume of wood products ranged between one to eight percent, according to the (FAO data 2017). Production of forest products has been healthiest in Asia-Pacific and North America due to a growing housing market (FAO, 2016). Other important factors in the wood products outlook include a decline in harvesting from natural forests and the emergence of planted forests as the major source of wood supply and technological developments such as increased plantation productivity through tree improvement, reduced wood requirements owing to expanded recycling, higher recovery, wider use of new composite products and production of cellulosic bio fuel (State of the World's Forests 2009). The world's forest plantation area, as reported to the Global Forest Resources Assessment 2005 (FAO, 2006), is 140.8 million hectares.

**Table 9. Growth Pattern of future demand of wood**

| Year | Demand (million m3) | Percentage increase average per annum |
|---|---|---|
| 2000 | 58.00 | - |
| 2005 | 74.00 | 5.52 |
| 2010 | 95.00 | 5.68 |
| 2015 | 123.00 | 5.89 |
| 2020 | 153.00 | 4.88 |

Source FAO 2009

**Indian paper industry:** The average annual turnover of the Indian paper industry is approximately Rs. 30,000 crores, which accounts for approximately 0.37% of the national GDP. It contributes Rs. 3000 crores to exchequer and provides direct employment opportunities to about 3.7lakh people and indirect employment of over 12lakh people. The industry employs wood, agro residues and recycled/ waste paper as the major raw material for manufacture. Indian paper industry ranks sixth among the energy intensive industries with an energy requirement of about 10 Mtpa of coal and 10.6 GWh of electricity. Table 10 shows the current status of Indian paper Industry. The medium sized agro based paper mills have the capacity from 30-350 tpd with a production share of 22% whereas the small waste paper based paper mills operate in the range10-500 tpd contributing to 47% of total country's production.

**Table 10. Status of the Indian Paper Industry**

| | |
|---|---|
| Number of mills | 759 |
| Installed capacity, Mt | 12.7 |
| Capacity utilization, % | ~ 90 |
| Production of Paper, Paperboard and Newsprint, Mtpa | 10.9 |
| Per capita Consumption (kg) | 9.3 |
| Annual Turnover, Rs. Crores | 30,000 |
| Contribution to Exchequer, Rs. Crores | 3000 |
| Employment Direct, million people | 0.37 |
| Indirect Employment, million people | 1.2 |
| Indian Share in World's Production, % | 2.6 |

Source: Working Group report of 12th FYP

## Structure of Indian Paper Industry

The Indian paper Industry has highly fragmented structure consisting of small, medium and large sized paper mills having capacities ranging from 10 to 1150 tpd employing wood, agro residues and recycled waste paper as major raw materials. The distribution of Indian paper industry based on the type of raw material used for making paper viz. wood, agro residues and recycled/waste paper is given in Table 11. Table 12 presents the category-wise production of paper from different raw materials in Indian paper Industry.

**Table 11. Structure of Indian Paper Industry**

| | No. of Mills | Production, Mtpa | Production Share (%) |
|---|---|---|---|
| Wood based (Large Integrated) | 30 | 3.40 | 31 |
| Agro based (Medium Scale) | 150 | 2.42 | 22 |
| Recycle Fibre based (Medium and Small Scale) | 579 | 5.10 | 47 |
| Total | 759 | 10.92 | 100 |

Source: Working Group report of 12th FYP

**Table 12. Production from Different Raw Materials**

| Variety | Raw Material | Production (Mtpa) | Total production (Mtpa) | % contribution |
|---|---|---|---|---|
| Writing Printing grade | Wood based | 2.53 | 4.18 | 38 |
| | Agro based | 0.78 | | |
| | Recycled fibre based | 0.87 | | |
| Packaging grade | Wood based | 0.84 | 5.90 | 54 |
| | Agro based | 1.64 | | |
| | Recycled fibre based | 3.43 | | |
| Newsprint grade | Wood based | 0.03 | 0.84 | 8 |
| | Agro based | Nil | | |
| | Recycled fibre based | 0.81 | | |
| Total Production | | | 10.92 | 100 |

The yearly production and consumption in Indian paper industry are presented in Table 13.

**Table 13: Yearly Production and Consumption in Indian Paper Industry**

| Year | Capacity, Mt | Production, Mt | Export, Mt | Import, Mt |
|---|---|---|---|---|
| 2005-06 | 7.32 | 6.80 | 0.312 | 0.981 |
| 2006-07 | 7.99 | 7.16 | 0.343 | 1.138 |
| 2007-08 | 8.32 | 7.33 | 0.329 | 1.341 |
| 2008-09 | 8.83 | 7.64 | 0.340 | 1.407 |
| 2009-10 | 9.34 | 8.02 | 0.382 | 1.464 |
| 2010-11 | 12.7 | 10.1 | 0.527 | 1.582 |
| 2011-12 | 13.55 | 10.9 | 0.545 | 2.336 |

Source: Working Group report of 12th FYP

**Natural fibre Industry:** Natural fibres are extracted from plants and are classified into three categories, depending on the part of the plant they are extracted from, e.g. Fruit fibres, Bast fibres and leaf fibres (Van Rijswijk, 2001) i.e.

Cotton, is world-wide one of the most important Fruit fibres used in the textile industry. Cotton is applied for the manufacturing of clothes, carpets, blankets, mobs and medical cotton wool. Coir (Coconut fibre) fibre is obtained from the husk of the fruit of the coconut palm, is used for brushes and mattresses, i.e.

Stem fibres: Jute; *Corchorus capsularis* and *Corchorus olitorius* are the main source of jute. The fibres are extracted from the ribbon of the stem. It is used as packaging material (bags), carpet backing, ropes, yarns and wall decoration. Flax Flax is used for the production of linen and canvas, ropes and sacks (Koussios, 2000), i.e.

Ramie: obtained from *Boehmeria nivea* can easily grow in tropical agroclimatic regions, is an expensive and durable fibre and can be dyed very easily, and is used in decorative fabrics than as construction material like in curtains, wallpaper, sewing thread and furniture covers, i.e.

Hemp: *Cannabis sativa* is used for the production of rope, fishing nets, paper, sacks, fire hoses and textile. Leaf fibres Sisal obtained from *Agave sisalana* is mainly used for ropes, mats, carpets and cement reinforcement (Van Rijswijk, 2001). In India many companies and industries are taking lands from small to large farmers for plantation of these fibre species and paying them money in return.

## Role of Small Scale Industries in Promoting Agro Forestry

It is estimated that each million rupees of investment in fixed assets in the small-scale sector leads to production of goods and/or services worth Rs.4.62 million annually, with an approximate value addition of ten percentage points. This sector also creates the largest employment opportunities outside agriculture. It is estimated that Rs.100,000 of investment in fixed assets in the small-scale sector generates employment for four people (Saigal and Bose, 2003). Though it is quite difficult to get an accurate overall picture of the total number of forestry enterprises in the country or their output due to lack of adequate information. However, a few estimates are available. The ICFRE has estimated that there are 4,235 wood-based factories in India (ICFRE 2001). Different small scale industries which are promoting industrial plantation by taking lands from the farmers for plantation and generating employment are:

**Sawmilling:** It is estimated that there are around 23,000 sawmills in the country and 98% of these are in the small category with annual log intake up to 3,000$m^3$. Annual production capacity is estimated at 27.12 million $m^3$ (GoI 1999). The survey conducted in the late 1970s indicated that 1,39,650 workers were employed in the sawmilling industry at the time (Tewari, 1995).

**Safety matches**: There are approximately 12,000 safety match making units in the country, and all except five are in the small-scale and cottage industries category. Over two-thirds of India's matches are produced in just two districts - Ramanathapuram and Tirunelveli, both in Tamil Nadu. The bulk of the wood comes from neighbouring Kerala, where there are over 400 small-scale units making veneers and splints for supply to the match industry. This ancillary industry employs over 15,000 people directly and indirectly and produces goods worth Rs. 150 million annually (Tandon 1991). Main raw material used are *Bombax ceiba, Evodia roxburghiana, Ailanthus malabarica, Hevea brazilensis.*

**Sports goods:** The main raw materials used are willow, cane, mulberry, maple, ash and rosewood (Tewari, 1995). According to projections made by the Forest Survey of India, the total wood demand for the sports goods industry was estimated to be 101,000 $m^3$ in 2000 (Tewari 1995).

**Katha and cutch**: Katha and cutch are products made from the heartwood of *Acacia catechu* tree. Katha is used as an ingredient of paan (betel) and paan masala chewing confectionery in India. Cutch is used for dying canvas and tanning leather. According to an estimate made a few years ago, 3,000 tonnes of katha was produced annually in India, of which 2,000 tonnes was produced in the factory sector, which also produced 4,500 tonnes of cutch. The total consumption of wood was estimated to be around 200,000m3 (round) (Tewari 1995).

**Lac**: Lac is produced from the secretions of a tiny insect *Laccifer lacca* Kerr., which is a parasite on a number of wild and cultivated plants. India is an important producer of lac and lac products. The present production of lac is about 15,000 metric tonnes, which is much less than the peak production of 91,199 metric tonnes achieved between 1978 and 1979. Lacquerware and lac turnery is a traditional industry based on lac. It is estimated that in Channapatna Taluka of Karnataka, over 35% of the workforce is engaged in lacquer work (Bahuguna and Shiva 2002).

**Bamboo and rattan products**: There are a number of SSFEs manufacturing bamboo and rattan based products. The main products manufactured from bamboo are handicraft items such as table mats, trays, lampshades and other household articles. Reed bamboo based traditional industries, such as mat and basket weaving, play a crucial role in the rural economy. Many tribes and ethinic groups (Bhanjaras, Bansforias, Kamars, Kotwalias, etc.) earn their living through bamboo handicraft work. Recently some bamboo mat board manufacturing units have also been established. It has been estimated that bamboo based SSFEs provide livelihoods to more than 300,000 village people in Kerala state alone (Rawat, 2001)

**Rattan (cane)**: Rattan extraction and utilization in India is, by and large, a cottage industry. It is estimated that there are around 2,000 small to medium sized rattan-based industrial units in India, employing over 200,000 people, some in the manufacture of a variety of handicraft items and furniture, and the rest predominantly in the rural areas in extraction, cleaning, processing and transportation. The rattan furniture industry produces goods worth Rs.50 million annually. About 10% of the goods are exported (Bhatnagar, 2007).

## Biofuel Industry in India

India's share of crude oil production is about 1 per cent of global crude oil production, whereas consumption amounts to 3.1 per cent of global consumption (Rapier, 2012). A number of private and Government organizations are involved in production and distribution of biofuel in India. The leaders in biofuel processing in India are, D1 Oil Plc, Reliance Industries Ltd, Godrej Agrovet, Emami Group, Aatmiya Biofuels Pvt Ltd., Gujarat Oelo Chem Limited (GOCL), Jain Irrigation System Ltd., Nova Bio Fuels Pvt. Ltd., Sagar *Jatropha* Oil Extractions Private Limited etc. It is wise to consider the oil yield potential of different edible and non-edible crops before selecting the crop as suitable source of biodiesel production (Tuli, 2011). Considering the food grain scarcity in developing countries like India, edible major crops may be spared as a potential source for bio-diesel production. Typical feed stocks for biodiesel production are soybean, canola/rapeseed, sunflower, cottonseed, palm seed and palm kernel, corn and mustard seed oil. Pork, beef and poultry fat and grease also can be converted to biodiesel. Palm oil and animal fat may have a high free fatty acid content, which causes soap formation that has adverse effects on downstream processing and leads to yield reduction (Bhardawaj *et.al.,* 2013). The detailed agricultural biofuel potential of India is estimated in Table 14.

**Table 14: Biofuel potential from agricultural crops in India.**

| Crop | Residue type | Prod. (tons) | RPR | Res. (dry wt.) (tons) | Sustain. Res. (20%) | Biochem. EtoH-low (litre) | Biochem EtoH-high (litre) |
|---|---|---|---|---|---|---|---|
| Rice | Straw/husk | 9.60E+07 | 1.8 | 1.73E+08 | 3.46E+07 | 3.80E+09 | 1.04E+10 |
| Wheat | Straw | 8.70E+07 | 1.6 | 1.39E+08 | 2.78E+07 | 3.10E+09 | 8.30E+09 |
| Jawar | Stalk | 7.00E+06 | 2.0 | 1.40E+07 | 2.80E+06 | 3.00E+08 | 8.00E+08 |
| Surgar cane | Bagasse/ leaves | 3.42E+08 | 0.4 | 1.37E+08 | 2.74E+07 | 3.00E+09 | 8.20E+09 |
| Bajra | Straw | 1.04E+07 | 2.0 | 2.07E+07 | 4.10E+06 | 5.00E+08 | 1.20E+09 |
| Maize | Stalk/cob | 2.17E+07 | 2.5 | 5.43E+07 | 1.09E+07 | 1.20E+09 | 3.30E+09 |
| Gram | Waste | 8.20E+06 | 1.6 | 1.32E+07 | 2.60E+06 | 3.00E+08 | 8.00E+08 |
| Tur (Arhar) | Shell/ waste | 2.90E+06 | 2.9 | 8.30E+06 | 1.70E+06 | 2.00E+08 | 5.00E+08 |
| Other cereal | Stalk | 4.60E+06 | 2 | 9.10E+06 | 1.80E+06 | 2.00E+08 | 5.00E+08 |
| Total | | | | | 1.36E+08 | 1.50E+10 | 4.10E+10 |

Source: (Swain 2014) Biofuel Production in India: Potential, Prospectus and Technology

## Profitable value chain model

An organized value chain model which eliminated the middle men and benefitted both the farmers and industries. The industries have been benefitted through quality and assured raw material which resulted in the industries venturing into certified wood thereby help to achieve forest certification resulting in global expert of paper from NAIP partner industries.

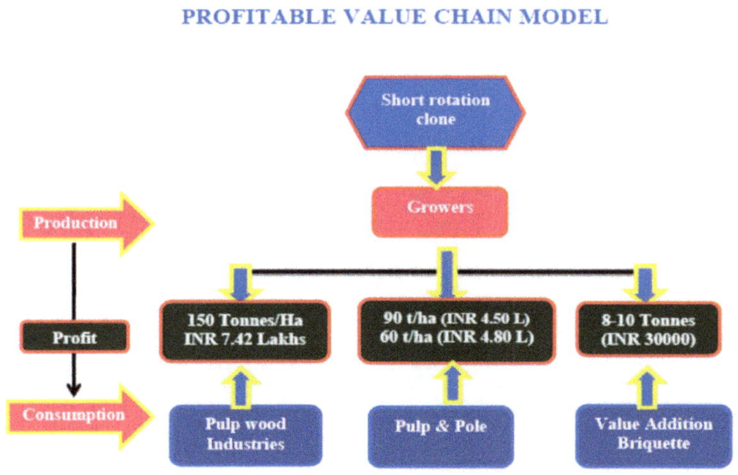


**Figure 35. Profitable chain model for farmers (Partiban et al, 2014)**

## Conclusions

Industrial agro forestry is a practice of agro forestry to meet the industrial requirement. The basic consideration in agro forestry is productivity, sustainability and adaptability. In many states of India many pulp and paper industries are approaching the farmers to cultivate pulp wood species with buy back agreement to meet the requirement of pulp and paper industry. In India mirrors global trends in small and medium enterprises, which indicate that small and medium enterprises represent one of the fastest growing industrial sectors in the world. The SSI sector accounts for around 95 per cent of the industrial units, 40 per cent of the manufacturing sector output, 45-50 per cent of exports (directly and through export houses etc), and provides direct employment to more than 19 million people in around 3.4 million registered SSI units. There is need for research to increase the system (trees and crops) productivity through low cost interventions and better techniques. Quality planting material, assured market and higher and assured returns and well established industry-farmer government linkage is the prime consideration for the rapid increase in industrial agro forestry. The government need to take confidence building measures and clearly formulate its policy and explain to farmers that their role is vitally important in the success of any agro industrial program me.

# Chapter 13

# Utilization of Agroforestry Trees

*Pawan Kumar Poonia, Pradip Kumar Sarkar*

## Introduction

In present days, forestry and agriculture sectors facing diverse challenges and constraints due to rising population pressure, urbanization, increasing food, pulp, fodder and timber requirement, coupled with natural resource degradation and global climate change. Therefore, it is indispensable to adopt corrective measures for improving the productivity of degraded forests (28 mha) and non-forest lands to safeguards to the existing natural resources. In this context, diversification of land use with agro-forestry can address some of these challenges. Agroforestry is considered as the future climate resilient land use system as it aims to increase the land productivity and protects the environment. Agroforestry or woody perennial based intercropping systems, has proved itself as a key component of sustainable agriculture and play important role in addressing the issues related to supply of wood, fuel and fodder and to preserve fragile agro-ecosystem. It not only arrest land degradation but also improve site productivity through interactions among trees, soil, crops and/or livestock, and thus restore parts, if not all, of the degraded lands (Kumar, 2006). Agroforestry, adds to the sustainability of agriculture and help in its diversification to attain higher benefits per unit land, when carefully selected fast growing multipurpose trees are raised with appropriate management (Johl *et al.*, 1986).

## Utilization of Agro-forestry Species

Agro-forestry utilization is the process of harvesting, converting and disposing of agro-forestry produce and other resources. Agro-forestry satisfies innumerable human needs, i.e. tangible or intangible and it cannot be bifurcated. Scientifically managed agro-forestry systems perform both needs, simultaneously, which makes it difficult to decide the priorities. The difficulty is pronounced in case of public forests wherein claims and needs of multiple beneficiaries are not easy to decide. At present agro-forestry meets almost half of the demand of fuel wood, 2/3 of the small timber, 70-80 per cent wood for plywood, 60 per cent raw material for paper pulp and 9-11% of the green fodder requirement of livestock, besides meeting the subsistence needs of households for food, fruit, fibre, medicine, timber etc (NRCAF, 2013). So we can promote the farmer to raise the tree on its farm and get maximum benefit from it, by efficient utilization of trees such as *Acacia, Cassia, Casuarina, Eucalyptus, Leucaena, Melia, Poplar* etc in different regions of the country. Some of basic utilization of different wood species is given in table-15.

**Table 15. Utilization of different woody species**

| Common Name | Scientific Name | Uses |
|---|---|---|
| Acacia | *Acacia auriculiformis* | Used for furniture and construction purposes. |
| Arjun | *Terminalia arjuna* | Agricultural implements, constructional, boat building, cart making and pit props. |
| Babul | *Acacia nilotica* | Agricultural implements, Constructional work for posts, beams, rafters, bridges etc |
| Bans | *Bambusa* spp. | Paper and wood composites |
| Beech | *Fagus* spp. | Furniture, flooring, musical instruments |
| Birch | *Betula* spp. | General purpose timber, plywood |
| Benteak | *Lagerstroemia microcarpa* | Construction, door and window frames, battens for tea- chests, packing cases, ammunition and explosive boxes, wooden ladders and trestles etc |
| Blue gum | *Eucalyptus* spp. | General purpose timber, also used for paper |
| Coconut | *Cocos nucifera* | As flooring, for furniture making, pulping, Constructional works. Local uses: manufacture of mats, ropes, brushes and baskets |
| Canarium | *Canarium* spp. | Light construction, furniture, veneer, plywood |
| Cottonwood | *Bombax* spp. | Plywood, block board, boxes and crates, furniture |
| Dhaman | *Grewia tiliifolia* | Agricultural implements, axe-helves, tool handles, constructional purposes, furniture, poles, shafts, panels and bent parts in carriage and spokes in cart, railway sleepers, boat and ship building, badminton rackets, balancing bench, hurdles, cricket stumps and bails; lorry and bus bodies and brush wares |
| Ebony | *Diospyros* spp. | Cutlery handles, musical instruments, craftwork |
| Fir, Douglas | *Pseudotsuga menziesii* | Plywood, structural work, sea defences |
| Fir, Himalayan | *Abies pindraw* | Building, light packing cases, fruit crates, joinery |
| Gamari | *Gmelina arborea* | Light construction, packaging, furniture |
| Gurjun | *Dipterocarpus* spp. | Strength applications in hidden work in furniture, building, and transport decking, plywood |
| Haldu | *Adina cordifolia* | Turnery and carving wood and is largely used for ply wood, Tea chests, furniture and cabinets, block board and tool handles |
| Indian elm | *Holoptelea integrifolia* | Constructional works, bobbins and cotton reels in textile mills, general purposes plywood, tea chests, decorative plywood, block boards, furniture, tool handles, agricultural implements, toys etc. |
| Kadam | *Anthocephalus chinensis* | Packing cases, ceiling boards, bottom of drawers, backing of cupboards, toys. |

| Common Name | Scientific Name | Uses |
|---|---|---|
| Keekar | *Prosopis juliflora* | fuel and charcoal |
| Mahogany | *Swietenia macrophylla* | Joinery, high quality furniture, plywood |
| Mango wood | *Mangifera indica* | Ceiling boards, window frames, plywood, furniture and cabinets, block boards, match splints and boxes, boat and ship buildings, wood articles, shoe-lasts |
| Meranti | *Shorea* spp. | All types of joinery, framing, doors and fittings. |
| Neem | *Azadirachta indica* | Light construction, furniture, doors and window frames boards and panels, agricultural implements, tool handles, musical, carving tools |
| Oak | *Quercus* spp. | Wide variety of applications, including joinery, furniture, fittings, flooring |
| Padauk | *Pterocarpus* spp. | Joinery, flooring, boat-building |
| Persimmon | *Diospyros virginiana* | Decorative ware, musical instruments, turnery |
| Pine | *Pinus* spp. | General utility work |
| Poplar | *Populus* spp. | Matches, turnery, minor craft goods |
| Red sanders | *Pterocarpus santalinus* | musical instruments especially "shamisen", carving idols, toys, ornamental house-posts and agricultural implements, sports goods, tool handles, verandah pillars, bent rims in cart construction,picture frames. |
| Rosewood | *Dalbergia* spp. | Furniture, musical instruments |
| Sissoo | *Dalbergia sissoo* | First class timber for cabinetry and furniture, door, window shutters and frames. |
| Silver oak | *Grevillea robusta* | Used for carpentry, cabinet work, joinery, furniture, parquet flooring, interior finishes, paneling, turnery and light construction work. |
| Teak | *Tectona grandis* | High quality furniture, joinery, garden furniture |
| Toon | *Toona ciliata* | Used for cabinets, furniture, plywood |
| Walnut | *Juglans* spp. | Furniture, joinery, gun butts |
| White mulberry | *Morus alba* | Hockey stick, cricket bats and other sports items |
| Willow | *Salix* spp. | Cricket bats, basket work, toys and other items |

The species (*e.g., Pongamia pinnata, Acacia catechu, Pterocarpus marsupium, Madhuca integrifolia, Azadirachta indica, Sapindus mukorossi, Emblica officinalis, Gmelina arborea, Dalbergia latifolia* and *Albizia procera*) traditionally grown by the farmers are very common practices in eastern region of India. Bakain (*Melia azedarach*), Teak (*Tectona grandis*) and Gamhar (*Gmelina arborea*) based agroforestry systems are found very common in the farm lands of different agro-ecological region of India and are even suitably planted for reclamation of degraded/ wastelands. Some of traditionally grown agroforestry species and their uses are listed in table 16.

**Table 16. Traditionally grown few agroforestry species in some parts of India and their uses (Sarkar *et al.*, 2017).**

| Species | Local name | Uses |
|---|---|---|
| *Slow growing* | | |
| *Azadirachta indica* A. Juss. | Neem | Pole for roof, fuelwood, small timber for furniture making, leaves as fodder, High medicinal value of all plant parts, leaves used as insecticide and pesticides to protect grains, roots and tuber crops in store conditions, seed oil and cake is used in cosmetics industries, biopesticides and as a manure in organic agriculture. |
| *Madhuca integrifolia* J. F. | Mahua | Timber is used in furniture making, leaves as fodder, all parts used as fuelwood, flower is edible food item for tribals and is also used for preparing local made alcoholic drinks, oil extracted from fruit is used as fuel oil and also in cosmetic industries, seed cakes obtained after extraction of oil constitute very good concentrated manure, several parts of the tree including the bark are used for medicinal properties, tribals of Jharkhand also consider tree as holy. |
| *Pongamia pinnata* (L.) Pierrre | Karanj | Branches and stems are used as fuelwood, oil extracted from fruit is used as fuel oil (i.e. biofuel) and also have medicinal value (skin injury) and seed cakes obtained after extraction of oil constitute a very good manure. |
| Medium growing | | |
| *Artocarpus lakoocha* Roxb. | Barhal/ Dahu | All parts are used as fuelwood, timber is used in furniture making, leaves and fuits are used as fodder, fruits are edible to people, heartwood is locally used as medicinal purpose for skin treatment |
| *Dalbergia latifolia* Roxb. | Kala shisham | Branches and stems are used as fuelwood, timber is used in furniture making, flooring and house construction, and leaves are used as fodder. |
| *Tectona grandis* Linn F. | Sagwan | Timber in furniture making, flooring and house construction, leaves and are used as fodder, bark is used as medicine for curing swelling and relieving skin irritation, drywood used as fuelwood. |
| *Terminalia arjuna* (Roxb. ex DC.) | Arjun | Branches and stems are used as fuelwood, timber for furniture making, flooring and house construction; all parts are used as medicine, barks used for heart diseases. |
| *Fast growing* | | |
| *Dalbergia sissoo* Roxb. | Shisham | Branches and stems are used as fuelwood, timber for furniture making, flooring and house construction, leaves are used as fodder, various parts are used as medicine |
| *Gmelina arborea* Roxb. | Gamhar | Branches and stems are used as fuelwood, small timber is used as pole, large timber for furniture making, flooring and house construction, leaves and fruits used as fodder. |
| *Melia azedarach* L. | Bakain | Used as a pole, stem and branches are used as fuelwood, small Timber is used for furniture making, leaves as fodder, all parts of tree used as medicinal/ pesticide/ insecticide purpose, various parts of the tree are used to protect stored grains, roots and tuber crops, *etc.* |

After wood, next important product from agro-forestry is fuel wood. The fuel wood potential of indigenous (*Acacia nilotica, Azadirachta indica, Casuarina equisetifolia, Dalbergia sissoo, Prosopis cineraria* and *Ziziphus mauritiana*) and exotic (*Acacia auriculiformis, A. tortilis, Eucalyptus, Prosopis juliflora*) trees revealed that calorific values range from 18.7 to 20.8 MJ kg$^{-1}$ and 17.3 to19.3 MJ kg$^{-1}$ respectively (NRCAF, 2013). Species such as *Casuarina equisetifolia, Prosopis juliflora, Leuceana leucocephala* have become prominent due to their potential for providing wood energy at the highest efficiency, shorter rotation and also their high adaptability to diverse habitats and climates. It may be one of the options to develop biomass based power plant to generate electricity. The fuel wood properties of some other potential agroforestry species are given in Table 17.

**Table 17. Fuelwood properties of some potential agroforestry species (Bhatt *et al.*, 2017)**

| *Species* | *Calorific values (MJ/kg)* | *density (g/cc)* | *Ash (%)* | *Moisture (%)* | *(FVI)* |
|---|---|---|---|---|---|
| *Albizia lucida* | 15.31 | 0.70±0.03 | 4.97±1.16 | 46.86±2.10 | 213.61 |
| *Flemingia semialata* | 16.50 | 0.66±0.05 | 3.52±1.12 | 54.42±0.84 | 307.61 |
| *Litsea polyantha* | 16.70 | 0.74±0.05 | 3.32±1.16 | 41.37±1.36 | 307.61 |
| *Mimusops elengi* | 16.56 | 0.80±0.09 | 3.98±1.14 | 46.35±1.01 | 229.90 |
| *Pongamia pinnata* | 16.68 | 0.79±0.005 | 3.55±1.30 | 47.76±0.84 | 330.97 |
| *Schleichera oleosa* | 16.09 | 0.86±0.08 | 5.85±1.09 | 47.07±1.01 | 229.90 |
| *Swietenia mahagoni* | 15.06 | 0.67±0.08 | 10.95±1.17 | 46.88±0.69 | 91.87 |

*Note: Fuelwood value Index (FVI)*

The other source of income from agro-forestry is NTFP's or minor forest products *i.e.*, the *Products of biological origin other than wood derived from forests, other wooded land and trees outside forests"*. The minor forest products such as natural gums & resins, lac cultivation, sericulture, apiculture, medicinal and aromatic plants in agro-forestry provides additional opportunity for earning livelihood. Some of basic products of different agro-forestry species are given in table- 18.

**Table 18. Non timber forest products of different agro-forestry species (Anon., 1976).**

| Sl. No. | Products | Examples |
|---|---|---|
| 1 | Aromatic plants | Palmarosa oil (extracted from *Cymbopogon martini* var. motia), Lemon grass oil (extracted from *Cymbopogon flexuosus*), Sandal oil (extracted from *Santalum album*), Khus oil (extracted from *Chrysopogon zizanioides*) |
| 2 | Bamboos | *Dendrocalamus strictus, Bambusa arundinacea,* Oxytenanthera spp., Ochlandra spp. |
| 3 | Bidi leaves | *Diospyros melanoxylon, D. Montana, Bauhinia racemosa, Castanopsis indica, Butea monosperma* |
| 4 | Beads for ornaments and decoration | *Abrus precatorius, Adenanthera pavonina, Corypha umbraculifera, Putranjiva roxburghii, Elaeocarpus* spp. |
| 5 | Canes | *Calamus rotang, C. viminalis, Daemonrops, Plectocomia* |
| 6 | Charcoal & Briquetting | Biomass of all trees |

| Sl. No. | Products | Examples |
|---|---|---|
| 7 | Cutch and Katha | Khair *(Acacia catechu), Acacia chundra* |
| 8 | Dye and colour yielding plants | *Butea monosperma*, Red sander, Khair, *Bixa orellana, C. sappan*, Mehandi |
| 9 | Edible plant products | *Mangifera indica, Syzygium cumini, Ziziphus mauritiana* |
| 10 | Fat oil yielding tress | *Aleurites fordii, Pongamia pinnata, Madhuca indica* |
| 11 | Fibre and floss yielding plants | *Sterculia villosa, Grewia tiliaefolia*, Anjan *(Hardwickia binata)*, Semal *(Bombax ceiba)*, Kapok *(Ceiba pentandra)* |
| 12 | Fodder and forage plants | Anjan, Babul, Subabul, Andropogon spp., Panicum spp. |
| 13 | Gum and resin exuding plants | Babul, Salai *(Boswellia serrata)*, Muttuga/Palash *(Butea monosperma)*, Honne *(Pterocarpus marsupium), Sterculia urens, Pinus* spp. *Canarium strictum, Vateria indica* |
| 14 | Leaves for plates, platters, bowls | *Butea monosperma, B. superba, Bauhinia vahlii*, Sal *(Shorea robusta)* |
| 15 | Medicinal plants | *Rauvolfia serpentina, Cinchona ledgeriana, Strychnos nux-vomica* |
| 16 | Saponin yielding plants and marking nut | *Sapindus mukorossi, S. trifoliatus, S. laurifolius, Acacia concina, Semecarpus anacardium* |
| 17 | Spices and condiments | *Cinnamomum zeylanicum, Curcuma aromatica, Piper nigrum*<br>Mango ginger *(Curcuma longa/C. amada)* |
| 18 | Tannin yielding plants | Babul *(Acacia nilotica)*, Wattle, Myrobalans (*Terminalia* spp.), Khair (*A. catechu*) |
| 19 | Others | Coir: *Cocos nucifera,*<br>Waxes: *Sapium sebiferum, Rhus succedanea*<br>Biopesticides: Chrysanthemum cinerariaefolium (Pyrethrum), *Derris ferruinea* (Rotenone) |
| **Animal products** | | |
| 20 | Honey and bees wax | *Apis dorsata, Apis indica* |
| 21 | Lac and shellac | *Laccifer lacca* (Kerria lacca), *Schleichera oleosa* (Kusumi), *Butea monosperma*, Ber, Ghont (Rangeeni) |
| 22 | Mulberry silk<br>Tassar silk<br>Muga silk<br>Eri silk (Assamese silk) | Bombyx mori: *Morus alba*<br>*Antheraea paphia* (Tassar) : *Terminalia* spp. Dindal<br>*A. assamensis* (Muga): *Litsea polyantha, Machilus* spp.<br>*Samia cynthia ricini*: Castor (*Ricinus cummunis*) |

For efficient utilization of NTFPs it is necessary to develop proper market and post-harvest management technologies. Similarly, fruit based industries linked with food supply chain and creation of infrastructural facility like cold storage and transport facilities have changed marketing of fruits, vegetables, medicinal and aromatic plants. Industries (wood based or crop based) can join hands together to process variety of products from agro-forestry system. Growing awareness towards valuation of environmental services and carbon trading will provide an opportunity for more plantations under agro-forestry systems on privately owned lands. There is a vast potential for investments and creation of employment in agro-forestry sector through involvement of corporate sector. Bio-fuel and bio-energy are yet other areas where industry can play an important role to meet the requirements for energy as well as for employment generation.

## Value Addition in Some important Agroforestry Species

In most of the cases, direct utilization of products derived from trees is not possible unless and until converting them into a desired shape, size, colour, odour and value added product as per the needs.

### Value addition in fruits

Fruits are generally perishable commodity and their wastages are very common due to improper handling, management, storage, distribution and marketing. On the other hand, these also play a vital role in generating employment and income and also improving nutrition and livelihood of rural and tribal masses. Now-a-days, various value added fruit products are getting more attracted towards consumers and are fetching higher prices and values. Value addition, thus is the process of changing or transforming a product from its original state to a more valuable state in order to avoid the glut and utilize the surplus during the off-season (Sharma *et al.*, 2014).

**Methods of processing for value addition in fruits-** Following are some of the methods:

a. **Innovative packaging:** Good quality package materials are necessary for long storage and disposal. Sometimes, packaging materials like wooden box or cartoon box are commonly used.

b. **Non thermal process:** It is used to pasteurise fruit juices for extending shelf life.

c. **Irradiation:** where fruit is exposed to ionised energy under controlled condition. It does not have any side effect. It helps in delaying in ripening. Example: Banana

d. **Heat processing:** This method is mainly used for dehydration of fruits (either by using oven or naturally by sun drying) for maintain very long shelf life.

e. **Refrigeration:** Storing of fruit pulp under very low temperature for further processing and value addition. It is common in converting mango pulp.

f. **Pre-packaging of fruits:** Now-a-days, wrapping of fruits is very common to fetch more market value and price.

g. **Wax emulsion or Coating for fruits:** It is very common to temperate fruits like apple for increasing shelf life, protects these from fungal infection, and reduces desiccation and weight loss during storage and disposal.

h. **Use of anti-fungal paste:** This is applied to some fruits like in Banana to reduce early spoiling.

The most common processed products those are being consumed by people are jam, ready to serve fruit drinks, chutneys, candies, pickles, squashes, concentrate, *etc*. Some of the value added fruits are mentioned below:

**Mango:** Various processed products like juice, pulp, concentrate, beverages, slices, dried form, frozen form, canned mango, amchur, mango leather, bar, jam, *etc*. are commonly available in the market.

**Litchi:** This fruit can be processed to prepare juice, beverages, candies/chocolate, etc.

**Jackfruit:** Various processed products like canned jackfruit bulbs in syrup, squash, raw jack pickle, roasted jack seeds, jack seed flour and candied jackfruit can be prepared from fruit (Srivastava *et al.*, 2017).

**Bael:** Various processed products such as nectar, squash, slab, toffee powder, *etc*. can be made with bael pulp.

**Jamun:** Jamun fruits can be processed into excellent quality fermented and non-fermented beverages. Besides that, good quality jelly, jam, leather can be prepared. A good quality jelly can also be prepared from its fruits. The processed seed powder is very useful in curing diabetes.

**Ber:** It can be processed to prepared murrabba, candy, dehydrated ber, pulp, jam, sharbat, squash and ready-to-serve beverage (Srivastava *et al.*, 2017) and powdered fruits after drying are also consumed.

**Aonla:** Various products like murabba, sweets, juice, squash, candy, powder, pickle, chutney, jam, *etc.* can be produced to maximize the market value.

**Tamarind:** The fruits can be processed into various products like jellies, chutneys, beverages (nectar, squash and syrup) Pickle, candy *etc.*

**Guava:** This fruit can be processed to have guava RTS, beverage, jam and jelly.

**Banana:** Few processed products commonly available in the market are pulp, beverage, powder and chips.

## Value addition to timbers

The timber gets more valued only when it is converted to some valuable products. For example, some species like *Eucalyptus, Casuarina sp.*, Teak, *etc.* are more demanding for pulp and paper industries and species like Poplar (*Populus deltoids*) is the major source for match box industries. Many of the agroforestry trees like Teak (*Tectona grandis*), Gamhar (*Gmelina arborea*), Shisham (*Dalbergia sissoo*), Arjun (*Terminalia arjuna*), Semal (*Bombax ceiba*) etc. are harvested and converted directly into various useful value added products *viz.*, furniture, door, windows, wooden frame, *etc.*.Various value added primary products developed from large scale industries include chips, lumber, veneer, plywood and their byproducts and secondary products derived from primary products include panels or engineered composites, cabinets, furniture and flooring, door, windows, etc.. Moreover, values of woods and wood products can be increased at various stages of processing. It can be done even by changing or enhancing the properties of wood and wood products. For example, making changes to dimensional stability or resistance to termites and decay in timbers or panels and producing more durable wood-based products can attract additional values and market prices. Thus, value added timber products industries can play major role in employment generation for the rural youth and economic development of the nation as well.

## Value addition in bamboos

Bamboos are known to be used as a substitute for hardwood. Various value-added bamboo products such as carpets, baskets, chair, tables, flower pots, construction beams, kitchen ware, latex-backed flooring on a roll, mouldings, traditional furniture, many home décor, packing materials, *etc.* are the major income source for many rural masses. Many species of bamboos are also highly exploited for large scale industries like paper making industries, fibre board industries, *etc.* Many converted products of bamboos like parquet flooring, veneer panels and laminated wood are found in the market. Conversion of bamboos into making incense sticks is very common to small scale industries.

The shoots of many of the bamboo species are edible and are highly being consumed in India because of their exotic taste, flavour, medicinal and nutritive value. Moreover, the edible shoots, having short shelf life are mostly available in the market during monsoon for a limited period in a year. Hence, to make these shoots available during offseason, they need to be processed and utilized immediately after harvesting. Various value added products like pickles, nuggets, dried shoots, powered form, papad, *etc.* can be made and have a scope in bamboos to develop more value added products. Some of the important edible bamboo species like *Dendrocalamus strictus, Bambusa bambos, B. tulda, D. asper, etc.* are commonly found to grow in many parts of India.

## Value addition in other NTFPs

Various value added products can be developed for increasing shelf life, available during offseason, gaining market value and generating high income. Following are some of the value added products:

**Lac:** Direct use of lac is not possible or acceptable unless it is converted to some other form. The various products like bangles, lac boxes, lac jewellery, home decors, *etc.* can be obtained.

**Gums, resins and latexes:** Commercial gums available in the market are in the form of dried exudations. These are also used in printing and finishing textiles, as a sizing for paper, in the paint and candy industries and as drugs. Resins are mainly obtained from species of different families like Anacardiaceae, Burseraceae, Dipterocarpaceae, Guttiferae, Hammamelidaceae, Leguminosae, Liliaceae, Pinaceae, Styracaceae and Umbelliferae. Resins are also used in medicine, for sizing paper, as a stiffening material for mats, in the preparation of sealing wax, incense and perfumes, *etc.* Various value added products of gums and resins are food additives, food thickener, *etc.* Latex of some species can be converted into rubber and some valuable products. Valuable products of natural latex rubber obtained from *Hevea brasiliensis* can be of rubber mats, rubber shoes, rubber belt, hand gloves, *etc.*

**Tans and dyes:** Tans and dyes are extracted from many tree species. For example, the processed products of cutch and katha obtained from *Acacia catechu* can be used for multiple purposes. It is also used in colour producing industries/ confectioneries.

**Fibre:** Fibre of many species like *Eucalyptus,* Teak, bamboos, *etc.* can be converted into and useful in making paper and other valuable products.

**Silk:** Value added products like silk sarees and highly commercial value of other products can be developed.

**Honey and wax:** Honey has attractive chemical properties for baking and a distinctive flavor when used as a sweetener and wax obtained from beehive is converted into candles, dolls, home decor, *etc.*

**Spices:** The bark powder of Cinnamon and Cassia can be used as spices.

**Medicinal:** Some species like barks of Arjun tree (*Terminalia arjuna*) and seeds of *T. chebula* and *T. bellerica* having medicinal value, can be converted to power form or juice or syrup form.

**Leaves:** Various products like Tendu (*Diospyros melanoxylon*) leaves for Bidi making, leaves of Sal (*Shorea robusta*) and Palash (*Butea monosperma*) for making plates/ cups, *etc.* can be prepared.

### Advantages of value addition

Following are the advantages of value addition to the products of agroforestry species:

a. Creation of product diversity
b. Income generation
c. Employment generation
d. Availability of products during offseason
e. Opportunities for others in gathering knowledge on value addition and improving upon already available value added products
f. Helps in reducing wastages

## Marketing Channels of Agroforestry Products

Wood and wood products are basic necessities and important entity in people life. Agroforestry provides us numerous materials like wood and non-wood products like gum and resins etc. The woody materials are mainly sold in firewood, wood-logs, or as standing trees. The non-wood forest products were gum, resin, leaves, flowers, forage, honey, silk and many other indigenous products, generally marketing system of non-wood forest products was non-commercialized. The tree products in villages were locally sold on cheep rate or gifted to others; even producer didn't know value because of imperfect marketing system. Major portion of tree produce was fire-wood or small pieces of wood, sold to bakers, restaurants and households while remaining products were timber, which serves as raw materials to different wood based industries like saw-mills, plywood and paper and pulp industries. After that, a user product like furniture, plywood is made. Therefore, marketing channel is required to get the good price of agroforestry produce. Marketing channels for agroforestry produce will bring the competitiveness in product prices which enable better prices for product. The centralized marketing channels for agroforestry products, which refer as producers/farmers to bring their products in market, like other agricultural products, products were sold directly to firms, saw-mill owners and other intermediaries.

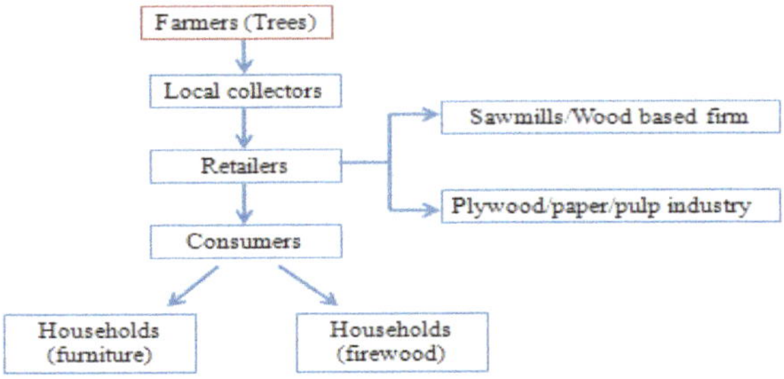


**Figure 38. Marketing channels of agroforestry products**

## Conclusions

Wood and wood products is an important entity in human life. The demands for woody raw material are bound to increase especially in the situations where the available wood substitutes are either expensive or pose environmental risks in production. Wood being renewable can meet such demands if used intelligently and success of any agro-forestry system, mainly depend upon quality planting material, marketing of agroforestry produce. Hence, tree improvement and production of quality planting material should get priority in future as well as to develop efficient wood processing techniques. Marketing system for agroforestry products was not in operation, and due to non commercialized marketing structure for wood and non-wood forest products, these products were either sold on the lower price at farm or available freely because of unawareness. Usually, farmers sold the whole trees to local assembler or collectors, which give lower price of wood logs, thus market exploitation situations were prevailed. This circumstance may lead bring down benefits to farmers and may discourage for tree plantation and adoption of agroforestry. Value addition in non-wood produces can help farmers in getting high remuneration. Research and development of value added products, acceptable to common man can change agriculture production scenario and consumer behaviour as well. Therefore, it is necessary to develop proper agro-forestry extension services, improve marketing channel along with value addition of non-wood produce to make agroforestry as a profitable business enterprises.

# Chapter 14

# Mechanization of Agroforestry

***Raj Kumar***

The demand of woody biomass both for industrial and energy use in India is estimated to increase markedly in the coming decades to satisfy the material and energy supply of bio-based economy. Moving towards more mechanized technology induces needs to invest in machinery, which may decrease the labour input/output volume ratio. In spite of an increase in revenue from timber sales the economic situation of many forestry enterprises is still, in many places, unsatisfactory. The greater supply required for new large saw mills and an increased demand for energy timber means that mechanization of operations in agroforestry is required. In Indian, most of the agroforestry operations are done manually that results in increased labour cost and time in all the agroforestry operations. In developed word, most of the agroforestry operations are mechanized that result in reduced labour requirement and time in agroforestry operations. The following is the brief description of important machinery used in agroforestry operations in developed countries.

**Figure 39. Mechanical planter**

**Mechanical planter**: This planting machine developed for Mondi by Anco Manufacturing of Piet Retief allows six planters to apply gel and plant simultaneously. The seedlings are carried in buckets by the planter, who drops them into a tube for planting and then tramps down the soil around the seedling with his feet. It is powered by a self-drive Fiori machine, and features hydraulic folding booms. It plants up an average of seven hectare per day at a rate of 40-50 plants per minute with six planters.

**Figure 40. Pitting machine**

**Pitting machine**: Jaaap (with mike) and Helgaard Steenkamp providing some information on the single head pitting machine from Novelquip Forestry. The machine averages 2000 pits/shift, and logs every pit made on a GPS so you can create a map of the compartment, making it easy to integrate silviculture with harvesting planning.

**Figure 41. Forwarder**

**Forwarder**: Forwarder is used for forwarding, and carrying bucked logs on the bed. Small-sized forwarders with grapple loaders, less than 3 tons in the loading weight, are mainly used in Japan.

**Figure 42. Harvester**

**Harvester**: Harvester is a machine for felling, delimbing and bucking. As to harvester, it is needed to buck accurately and delimb thick branch, as well as processor. One-grip type model of harvester has a majority because excavators are used as base machine in most cases.

**Processor**: Processor is a machine for delimbing and bucking, most utilized high-performance forestry machine. Processor is required abilities to measure the length of logs exactly and delimb thick branch of Hinoki (*Chamaecyparis obtusa*) which is one of the main wood species.

**De-stumper**: Like a giant dentist this highly effective head extracts stumps up to 600 mm diameter by cutting them off 80 mm below ground level. The DS800 de-stumper is manufactured by Ralf Conrad of Rupa Timber & Products, Piet Retief. The machine averages 60 stumps/hour. Stumps can be used by the operator to make charcoal. It has been observed that stumps constitute around 8% of the total timber extractable from a compartment. The de-stumper leaves the roots behind in the soil. Ralf says he makes one ton of charcoal from 6-8 tons stumps.

**Figure 43. Processor**

**MECHANIZED TASK FORCE:** A mechanized, or heavy equipment task force is made up of multiple forestry machines, each different and working in concert. The machine combination varies by the tasks at-hand, and requires operators that understand their individual role in achieving the overall objective, i.e. planting, pruning, tree felling, tree removal (skidding), and transportation. Mechanization has its own constraints. Machines must be used on sites that are within the machine's design parameters and physical operating limits to safely fell and handle trees, i.e. ground slope, ground pressure and tree size. Consistently safer operations are realized when workers are properly trained. Mechanized Task Force training must be appropriate for specific terrain, vegetation, and operational objectives. Supporting personnel must train with the equipment to learn the essentials for maximizing efficiency and safety in an inherently dangerous job. To understand the size and forces of large tree felling operations training must be live and in-woods, and include personnel that learn to effectively maximize advantages of time, space, and purpose.

**Figure 44. De-stumper**

## Conclusion

The most recent research on the forest machines development needs focuses on forest energy harvesting. This is natural, because compared with well-established roundwood harvesting systems the forest energy machinery and the whole supply chain is still under development. In India, the high cost of cutting, unskilled operator and non-availability of skilled labour were found the main challenges in agroforestry mechanization. Therefore to make agroforestry a success, mechanization and availability of skilled labour is utmost important in 21$^{st}$ century.

# Chapter 15

# National Agroforestry Programs and Policy

*Kanica Chauhan, Sneha Dobhal*

## Introduction

AGROFORESTRY – a collective name for land-use systems and technologies where woody perennials (trees, shrubs, palms, bamboos, etc.) are deliberately used on the same land-management units as agricultural crops and/or animals, in some form of spatial arrangement or temporal sequence. In agroforestry systems, there are both ecological and economical interactions between the different components (Lundgren and Raintree, 1982). Traditionally, local people preferred agroforestry practices for the inter-dependent benefits of the three components, viz. trees, crops and livestock in addition to the 6Fs, i.e. food, fruit, fodder, fuel, fertilizer and fibre.

Agroforestry has significant potential to generate employment for rural and urban population. It has also been estimated that about 65 % of the country's timber requirement is met from the agroforestry practices. It is also recognized that agroforestry is perhaps the only alternative to meet the target of increasing forest or tree cover to 33 per cent from the present level of less than 25 per cent, as envisaged in the National Forest Policy (1988).

Agroforestry can practically mitigate the climate change effects through microclimate moderation and natural resources conservation in the short run and through carbon sequestration in the long run. Agroforestry systems provide significant rescue to the present challenges of food, nutrition, energy, employment and environmental security. However, appropriate research interventions, adequate investment, suitable extension strategies, incentives to agroforestry practitioners, enabling legal and regulatory environment, marketing of agroforestry produce, post-harvest processing, development of new products, and above all a forward looking National Agroforestry Policy is required to address these issues.

Larger parts of the country are still dependant on rainfall for farming and their livelihood. Due to the changing climatic pattern, rainfall is becoming more erratic, making cultivation a high risk and less productive profession over the years. It has therefore, becoming increasingly difficult for the majority of the Indian farmers to sustain their farm production, productivity and income. In order to make agriculture less vulnerable to climatic aberrations, Government of India formulated the National Agroforestry Policy in 2014. The Agroforestry Policy has identified various factors those have impeded the growth of the sector like non availability of quality planting material, prohibitive legal frame work particularly for felling and transit regulations, inadequate extension mechanism, lack of proper institutional support and focused programme for its development etc. The policy recommends for setting up of a Mission or Board to address development of agroforestry sector in an organised manner. The Sub-Mission on Agroforestry (SMAF) under NMSA (National mission on sustainable agriculture) is an initiative to combat these problems. The aim of the submission is to expand the tree coverage on farmland in complementary with agricultural crops.

## National Agroforestry Mission

### Mission Objectives

The National Agroforestry Mission was established with the following objectives:

1. To encourage and expand tree plantation in complementary and integrated manner with crops and livestock to improve productivity, employment opportunities, income generation and livelihoods of rural households, especially the small farmers.
2. To ensure availability of quality planting material like seeds, seedlings, clones, hybrids, improved varieties, etc.
3. To popularize various Agroforestry practices/models suitable to different agro ecological regions and land use conditions.
4. To create database, information and knowledge support in the area of agroforestry.
5. To provide extension and capacity building support to agroforestry sector.

### Mission Strategy

The multi-pronged strategies to achieve these objectives for enhancing the growth of agroforestry in India included:

1. Expanding the coverage under tree plantation in arable land suitable to local agro climatic and land use conditions to provide livelihood, environmental and bio diversity protection by encouraging farmers to grow trees in their farmland along with crops/cropping systems and/or livestock as an integral component of farming system.
2. Promoting setting up of new small nurseries and hi-tech big nurseries for producing quality planting materials like seeds, seedlings, clones, improved varieties to meet the requirement of quality planting material / seeds for the farmers.

3. Promoting various Agroforestry practices/models suitable to different agro ecological regions and land use conditions that will support adaptation and mitigation efforts in climate change. Promoting sustainable Agrisilvicultural systems, Silvipastoral systems, Agrisilvopastoral system, other systems of agroforestry viz. Apiculture with Trees, Aqua forestry etc.
4. Promoting Peripheral and Boundary Plantation on farms will serve as fencing of farm, demarcation of farm boundary, stabilizing farm bonds, protecting from soil erosion, improving soil moisture, enrichment of soil organic matter without affecting coverage under crops.
5. Low Density Plantation on Farm Lands including intermediate/strip plantation, High Density Block Plantation will also be supported in complementarily with crops/cropping system as agroforestry system.
6. Creating database on area under agroforestry, status of soil organic carbon, information and knowledge support etc. with use of ICT.
7. Capacity building / training for Development officials/workers, scientists and farmers under National & International exchange programmes; extension activities including Demonstration of suitable Agrisilvicultural, Silvipastoral, agrisilvopastoral system etc., Seminars/workshops / conferences/fairs & exhibitions and exposure visits at National and International level.
8. Implementation of the scheme only in the states having liberalized transit regulations for transport of timber and will be extended to other states as and when such relaxations are notified by them. Liberal transit rules will be a precondition for availing the benefit of the programmme.
9. The state Govt has to decide the Department / Agency for implementation of the progamme. Institutional mechanism as envisaged in the NMSA Guidelines would be adopted for the sub-mission being a submission of NMSA with the change that the implementing department / agency will be designated as Member Secretary.
10. Endemic and other species including trees of medicinal value suitable to the agroclimatic conditions will be promoted under the programme. Any species which are exotic, not suitable to the agro-ecology and not part of approved afforestation programme will not be supported.
11. Soil Health Cards will be made a pre-requisite for farmers getting the benefit under the programme to indicate the soil carbon status and facilitating assessment of Soil Carbon improvement from time to time.
12. The choice of stock type (bare-root or container or poly bags) used may depend on species, local availability, planting method preferred or cost. It is important that the seedlings for plantation should be of the highest quality possible. The shoot and root systems of the stock must be large enough and in balance so that the seedlings have a good probability of establishing and competing successfully in the field. A good fibrous/tape root system is desirable for most species. Seedlings should be free from diseases, and

appear healthy. In particular, checking the quality of the roots system will be emphasized. It is important to be satisfied that the plants conform to Forest/Horticulture Department's regulations or, any other National/State regulations on quality.

13. Physical verification of the plantation will be done by the competent authority of State Government and a verification certificate needs to be obtained by each beneficiary.

## Mission Interventions

The Sub-Mission has following broad interventions:

1. Nursery Development for quality planting material (NDQPM): Assistance shall be given for nurseries (small, big & hi-tech nurseries) for producing quality planting material to meet the requirement of planting material.
2. Peripheral and Boundary Plantation(PBP) : To make potential use of the area occupied by these bunds around the periphery of the farmers fields, tree species can be grown as peripheral/boundary plantations to add more income to the farmers' basket. This will not only make effective use of the precious land for livelihood support but also for generating additional income opportunities to the farmers. It will also help in stabilizing the bunds and reducing soil erosion.
3. Low Density Plantation on Farm Lands(LDPFL) : Low Density Block Plantation(HDBP) ranging from more than 100 plants/ha to more than 500 plants/ha without sacrificing the yield of the existing crops/cropping systems, shall be incentivized at the proportionate rates as applicable to per plant expenditure.
4. High Density Block Plantation (HDBP): High density Block Plantations on farm lands ( HDBP) will be supported as a complementary source of income to the farmers. Differential planting densities ranging from more than 500 plants/ha to 1500 plants/ha as intermediate blocks / strip plantations /wind breaks would be supported. Farmers can take up block plantation in waste and degraded land not suitable for growing crops to make productive use of these lands in creating livelihood and income opportunities for them. In addition, the trees will help in enriching the soil and making it fertile & more productive thereby bringing land under crops in times to come.
5. Capacity Building & Trainings: Capacity Building and Training will be one of the important interventions of the sub-mission. Activities like training of farmers/field workers with a view to ensure growing quality planting material, skill development, awareness campaign, Publications, Information sharing, exposure visits at national and international levels, seminars/ Workshops at national/international levels etc. would be supported under this. States can utilize up to 5% of the allocated funds for capacity building and training activities.
6. Demonstration of Agroforestry Models: Specific projects for demonstration to bridge the yield gap and for extension purposes on area specific

innovative agroforestry models by SAUs / ICAR Institutes/ CAUs/CSIR/ ICFRE institutes / State Government/ other National & International level agencies/ organizations will be supported particularly in the field of climate resilient agroforestry system & studies on carbon foot print, carbon sequestration, nitrogen fixation etc., with the approval of Project Sanctioning Committee on case to case basis.

## National Bamboo Mission

Mission Bamboo is a versatile group of plants which is capable of providing ecological, economic and livelihood security to the people. Till recently, it has remained confined to the forests (12.8% of forest cover); two third of the growing stock located in the North-Eastern States. Importance of this crop as a source of raw material for industrial and domestic use with its growing demand all over the country necessitated its cultivation in farm lands and in degraded lands. With a view to harness the potential of bamboo crop, Department of Agriculture & Cooperation (DAC), Ministry of Agriculture & Farmers Welfare is implementing a 100% Centrally Sponsored Scheme called Mission for Integrated Development of Horticulture (MIDH) in which National Bamboo Mission (NBM) is being implemented as a sub scheme. The Mission envisages promoting holistic growth of bamboo sector by adopting area-based, regionally differentiated strategy and to increase the area under bamboo cultivation and marketing. Under the Mission, steps have been taken to increase the availability of quality planting material by supporting the setting up of new nurseries and strengthening of existing ones. To address forward integration, the Mission is taking steps to strengthen marketing of bamboo products, especially those of handicraft items.

## Objectives

1. To promote the growth of the bamboo sector through as an area based regionally differentiated strategy;
2. To increase the coverage of area under bamboo in potential areas with improved varieties to enhance yield;
3. To promote marketing of bamboo and bamboo based handicrafts;
4. To establish convergence and synergy among stake holders for the development of bamboo;
5. To promote, develop and disseminate technologies through a seamless blend of traditional wisdom and modern scientific knowledge;
6. To generate employment opportunities for skilled and unskilled persons, especially unemployed youths.

States/ districts covered: The Mission will focus on development of bamboo in limited States where it has social, commercial and economical advantage, particularly in the North Eastern region and States including Madhya Pradesh, Maharashtra, Chhattisgarh, Odisha, Karnataka, Uttarakhand, Bihar, Jharkhand, Andhra Pradesh, Telangana, Gujarat, Tamil Nadu and Kerala. The Mission is expected to establish about 4000 treatment/ product development units and bring more than 100000 ha area under plantation during the period 2018-19 & 2019-20.

## Strategy

To achieve the above objectives, the Mission would adopt the following strategies:

- Adopt a coordinated approach covering production and marketing to assure appropriate returns to growers/producers.
- Promote Research and Development (R&D) of varieties and technologies for enhanced production.
- Enhance acreage (in forest and non-forest areas) and productivity of bamboo through varietal change and improved agriculture practice.
- Promote partnership, convergence and synergy among R&D and marketing agencies in public as well as private sectors, at all levels.
- Promote where appropriate, cooperatives and self-help groups to ensure support and adequate returns to farmers.
- To generate employment opportunities for skilled and unskilled persons, especially unemployed youths.
- Set up National, State and sub-State level structures, to ensure adequate returns for the produce of the farmers and eliminate middlemen, to the extent possible.

## Key elements of National Bamboo Mission

- Research and Development
- Plantation infrastructure development
- Production of Planting Material
- Area expansion under Bamboo
- Improvement of Existing Stock
- Technology Transfer & HRD
- Pest and disease management of bamboo
- Creation of Water resources
- Innovative Interventions
- Post harvest storage and treatment facilities for bamboo
- Establishment of marketing infrastructure

# National Agroforestry Policy 2014

National Agroforestry Policy, 2014 has been formulated with the objective to bring coordination, convergence and synergy between various elements of agroforestry scattered in various existing Missions, programmes and schemes of the Government. The policy will provide a platform to stimulate the growth of agroforestry in India. It will encourage farmers for agroforestry as an integral component of farming system which will lead to meet the ever increasing demand of timber, food, fuel, fodder, fertilizer, fiber, and other agroforestry products;

conserving the natural resources and forest; protecting the environment & providing environmental security; and increasing the forest / tree cover. A major role for agroforestry is emerging in the domain of environmental services. Agroforestry is known to have the potential to mitigate the climate change effects through microclimate moderation and natural resources conservation in the short run and through carbon sequestration in the long run. Agroforestry species are known to sequester as much carbon in below ground biomass as the primary forests, and far greater than the crop and grass systems.

## Preamble

Agroforestry is defined as a land use system which integrates trees and shrubs on farmlands and rural landscapes to enhance productivity, profitability, diversity and ecosystem sustainability. It is a dynamic, ecologically based, natural resource management system that, through integration of woody perennials on farms and in the agricultural landscape, diversifies and sustains production and builds social institutions.

## Need for Agroforestry Policy in India

- Absence of a dedicated and focused national policy and a suitable institutional mechanism
- Lack of an integrated farming systems approach
- Restrictive regulatory regime
- Inadequate attempts at liberalization of restrictive regulations
- Insufficient research, extension and capacity building
- Dearth of quality planting material
- Institutional finance and insurance coverage
- Weak market access for agroforestry produce
- Industry operations at a sub-optimal level

## Goal

The major policy goals are:

- Setting up a National Agroforestry Mission or an Agro forestry Board to implement the National Policy by bringing coordination, convergence and synergy among various elements of agroforestry scattered in various existing, missions, programmes, schemes and agencies pertaining to agriculture, environment, forestry, and rural development sectors of the Government.
- Improving the productivity; employment, income and livelihood opportunities of rural households, especially of the smallholder farmers through agroforestry.
- Meeting the ever increasing demand of timber, food, fuel, fodder, fertilizer, fibre, and other agroforestry products; conserving the natural resources and

forest; protecting the environment & providing environmental security; and increasing the forest / tree cover, there is a need to increase the availability of these from outside the natural forests.

## Basic Objectives

The basic objectives of the National Agroforestry Policy are to:

- Encourage and expand tree plantation in complementarily and integrated manner with crops and livestock to improve productivity, employment, income and livelihoods of rural households, especially the small holder farmers.
- Protect and stabilize ecosystems, and promote resilient cropping and farming systems to minimize the risk during extreme climatic events.
- Meet the raw material requirements of wood based industries and reduce import of wood and wood products to save foreign exchange.
- Supplement the availability of agroforestry products (AFPs), such as the fuel-wood, fodder, non-timber forest produce and small timber of the rural and tribal populations, thereby reducing the pressure on existing forests.
- Complement achieving the target of increasing forest/tree cover to promote ecological stability, especially in the vulnerable regions.
- Develop capacity and strengthen research in agroforestry and create a massive people's movement for achieving these objectives and to minimize pressure on existing forests

## Strategy

- ❑ Establishment of institutional setup at national level to promote Agroforestry
- ❑ Simple regulatory mechanism
- ❑ Development of a sound database & information system
- ❑ Investing in research, extension and capacity building and related services
- ❑ Improving famers' access to quality planting material
- ❑ Providing institutional credit and insurance cover for agroforestry
- ❑ Facilitating increased participation of industries dealing with agroforestry produce
- ❑ Strengthening farmer access to markets for tree products
- ❑ Incentives to farmers for adopting agroforestry
- ❑ Promoting sustainable agroforestry for renewable biomass based energy

# References


Agarwal NK, *et al.* 2017. Promotion of Short Rotation Agro and Farm Forestry System in Gujarat and Maharashtra by JK Paper Limited Unit-Cpm. *Indian Forester,* 143(13): 862-868.

Ahmad F, *et al.* 2018. *Agroforest Syst.* https://doi.org/10.1007/s10457-018-0233-7

Alao JS and Shuaibu RB. 2013. Agroforestry practices and concepts in sustainable land use systems in Nigeria. *Journal of Horticulture and Forestry,* 5(10), 156-159.

Albrecht A and Kandji ST. 2003. Carbon sequestration in tropical agroforestry systems. *Agriculture, Ecosystems & Environment,* 99:15-27.

Allen S *et al.* 2004. Safety net role of tree roots: experimental evidence from an alley cropping system. *For Ecol Manage,* 192:395–407.

Alston JM, *et al.* 1995. Science under scarcity; Principle & Practice for Agriculture research & priority setting. *Cornell University Press.* pp-618.

Ponzetta MP. 2016. The sustainability of wildlife in Agroforestry land. *Agriculture and Agriculture Science Procedia,* 8, 148-157.

Andrew G and Masozera M. 2010. Payment for ecosystem services and poverty reduction in Rwanda. *Journal of Sustainable Development in Africa,* 12 (3).

Anonymous. 1976. Indian forest utilization, Volume-I & II. ICFRE Publication, Dehradun.

Anonymous. 2000. Proceedings of the symposium on "Statistical methodology for agroforestry research ". *Journal of Indian Society of Agricultural Statistics* 53(1): 82-90

Arico S *et al.* 2005. Ecosystems and human well-being. Technical report, Millennium Ecosystem Assessment Board.

Bahuguna R and Shiva MP. 2002. Lac: A review on present status and its prospects. *Usuf. Mgt,* 3 (1&2): 63-70.

Barrios E. 2007. Soil biota, ecosystem services and land productivity. *Ecological Economics*, 64: 269–85.

Barrios, E., *et al.* 1997. Light fraction soil organic matter and available nitrogen following trees and maize. *Soil Science Society of America Journal*, 61(3): 826–31.

Barros, E., *et al.* 2008. Comunidades da macrofauna do solo daAmaz˜oniaBrasileira. *In* F. M. S. Moreira, J. O. Siqueira, and L.Brussaard, (Eds) Biodiversidade do Solo emEcossistemasBrasileiros. Editora UFLA, Minas Gerais,Brazil, pp. 171–191.

Benjamin TJ *et al.* 2000. Defining competition vectors in a temperate alley cropping system in the midwestern USA. *Agroforestry Systems*, 48:79-93.

Bhardawaj A *et al.* 2013. A Review of Biofuel Policy in India: Current Status and Future Perspectives. *Int J Appl Eng Res*, 8: 1907-1912.

Bhatnagar P, *et al.* 2007). *Sustainable Forestry*, Vol.I,(2).

Bhatt BP, *et al.* 2017. Fuelwood characteristics of some important trees and shrubs and emission of carbon dioxide in different states of eastern India. Energy Sources, Part A: Recovery, Utilization and environmental effects, 39(4): 414-418.

Bisht V, *et al.* 2015. Changes in soil chemical properties under Eucalyptus (*Eucalyptus tereticornis*) based agroforestry systems in Semi-Arid region of Haryana. *International Journal of Tropical Agriculture*, 33(2): 1257-1260.

Box GEP, *et al.* 1998. *Statistics for experimenters.* John Wiley, New York.

Boyd J and Banzhaf S. 2007. What are ecosystem services? The need for standardized environmental accounting units. *Ecol Econ*, 63: 616–626.

Boyd J and Banzhaf S. 2007. What are ecosystem services? The need for standardized environmental accounting units. *Ecological Economics*, 63(2-3): 616-626.

Branca G *et al.* 2013. Food security, climate change, and sustainable land management. A review. *Agronomy for Sustainable Development*. doi:10.1007/s13593-013-0133-1.

Brevik EC. 2009. Soil health and productivity. In: Soils, Plant Growth and Crop Production. W. Verheye (Ed.). Encyclopaedia of Life Support Systems (EOLSS), Developed under the Auspices of the UNESCO, EOLSS Publishers, Oxford, UK. http://www.eolss.net.

Bucheli VJP and Bokelmann W. 2017. Agroforestry systems for biodiversity and ecosystem services: the case of the Sibundoy Valley in the Colombian province of Putumayo. *International Journal of Biodiversity Science, Ecosystem Services & Management*, 13(1): 380-397, DOI: 10.1080/21513732.2017.1391879

Buresh RJ and Tian G. 1998. Soil improvement by trees in sub-Saharan Africa. *Agroforestry Systems*, 38: 51–76.

Burgess PJ. 1999. Effects of agroforestry on farm biodiversity in the UK. *Scottish Forestry*, 53(1): 24-27.

Burgess PJ *et al.* 2003. The impact of silvoarable agroforestry with poplar on farm profitability and biological diversity: Final report to Defra Project AF0105.

Cannell MGR *et al.* 1996. The central agroforestry hypothesis: The trees must acquire resources that the crop would not otherwise acquire. *Agroforestry Systems*, 34: 27-31.

Cassman KG. 1999. Ecological intensification of cereal production systems: yield potential, soil quality, and precision agriculture. *Proc Natl Acad Sci USA*, 96: 5952–5959.

Castro A, *et al.* 2009. Quesungual slash and mulch agroforestry system (QSMAS): Improving crop water productivity, food security and resource quality in the subhumid tropics. Colombo, Sri Lanka: CGIAR Challenge Program on Water and Food.

Chaudhari, SK. 2016. Soil health in India: retrospective and perspective. *Bulletin of the Indian Society of Soil Science*, 30: 34-52.

Chauhan HS. 2000. Performance of Poplar (*Populus deltoides*) based agroforestry system using aromatic crops. *Indian Journal of Agroforestry*, 2: 17-21.

Chundawat BS and Gautam SK. 1993. Inrtoduction. **In**: Textbook of agroforestry. Oxford and IBH Publishing Co. Pvt. Ltd., New Delhi, India.

Costanza R *et al.* 1997. The value of the world's ecosystem services and natural capital. *Nature*, 387: 253-260.

Crowe SR and McAdam J. 1993. Factors affecting herbage biomass production in a mature tree silvopastoral system. *Agroforestry Forum*, 4(3): 14-18.

Cuthbertson A and McAdam J. 1996. The effect of tree density and species in carabid beetles in a range of pasture-tree agroforestry systems on a lowland site. *Agroforestry Forum*, 7(3): 17-20.

Daily GC. 1997. Introduction: What are Ecosystem Services? In: Daily, G.C. (Ed.) Nature's Services: Societal Dependence on Natural Ecosystems (pp1-10). Island Press, Washington, D.C.

De Groot *et al.* 2002. A typology for the classification, description and valuation of ecosystem functions, goods and services. *Ecological Economics*, 41:393–408.

Dhanya B *et al.* 2014. Ecosystem Services & Management. *International Journal of Biodiversity Science*,10 (2): 101–111.

Dhillon A *et al.* 2001. An economic analysis of poplar cultivation. *Indian Forester*, 127(1): 86-90.

Dickmann DI and Keathley DE. 1996. Linkage physiology and molecular genetics on *Populus* ideotype. In: Biology of *Populus* (Eds. R. F. Stettler, H. D. Bradshaw Jr., Heilman and T. M. Hinckley). NRC Research Press, Ottawa, Canada, 468- 490pp.

Dixon RK. 1995. Agroforestry system: sources or sinks of greenhouse gases? *Agroforest Syst*, 31, 99–116.

Dobo B *et al.* 2018. Effect of tree-enset-coffee based agro-forestry practices on arbuscularmycorrhizal fungi (AMF) species diversity and spore density. *Agroforestry System*, 92 (2): 525–540.

Dollinger J and Jose S. 2018. Agroforestry for soil health. *Agroforestry Systems*, 92 (2): 213-219. https://doi.org/ 10.1007/s10457-018-0223-9.

Doran JW and Parkin TB. 1994. Defining and assessing soil quality. In: Defining Soil Quality for a Sustainable Environment. Doran, J. W., Coleman, D. C., Bezdicek, D. F., and Stewart, B. A. (eds.). Soil Science Society of America, Inc., Madison, pp. 3-21.

Doran JW *et al.* 1996. Soil health and sustainability. In: Sparks, D.L. (ed.) Advances in agronomy, vol.56. Academic Press, New York.

Duchhart I *et al.* 1989. Planning methods for agroforestry. *Agroforestry System.* 7, 227-258.

Durairasu P and Parthiban KT. 2014. Wood based Industries, In: Industrial Agroforestry: perspectives and prospective. Scientific Publisher; Jodhpur; New Delhi: India. p1-3.

Ehrlich PR and Mooney HA. 1983. Extinction, substitution, and ecosystem services. *Bioscience*, 33: 248–254.

Ellis F. 2000. Rural livelihoods and diversity in developing countries, Oxford University Press, Oxford.

Fall D *et al.* 2012. Effect of distance and depth on microbial biomass and mineral nitrogen content under Acacia senegal (L.) Willd.trees. *Journal of Environmental Management*, 95: 60-64.

FAO (Food and Agriculture Organisation of the United Nations). (1986). Small Scale Forest Based Processing Enterprises. Rome: FAO.

Fisher B and Turner K. 2008. Ecosystem services: Classification for valuation. *Biological Conservation* 141: 1167-1169.

FAO (Food and Agriculture Organisation) Rome. 2016. Status of forest based industries.

Frank E and Eduardo S. 2003. Biomass dynamics of *Erythrina lanceolata* as influenced by shoot pruning intensity in Costa Rica. *Agroforestry System*, 57(1):19-28.

FSI. 2015. State of Forest Report 2015. Forest Survey of India. Ministry of Environment and Forests, DehraDun.

FSI. State of Forest Report 1999, Forest Survey of India, Dehra Dun.2000; 8-16pp

Garrett HE and McGraw RL. 2000. Alley cropping practices. In H.E. Garrett., Rietveld, W.J., Fisher, R.F. (Eds), North American agroforestry: an integrated science and practice. ASA, Madison, pp 149–188.

Garrity DP. 2004. Agroforestry and the achievement of the millennium development goals. *Agrofor Syst*, 61: 5–17.

Garrity DP. 2004. Agroforestry and the achievement of the Millenium Development Goals. *Agroforestry Systems*, 61: 5-17.

Ghose PK *et al.* 2008. Watershed based integrated resource management in hill ecosystem- A sustainable approach for livelihood improvement, ICAR Research Complex for NEH Region, Umiam Meghalaya, 1-14.

Gold M *et al.* 2013. Handbook for Agroforestry Planning and Design, The Center for Agroforestry, University of Missouri, Missouri, United States, 83p.

Gold MA, Garrett HE. 2009. Agroforestry nomenclature, concepts, and practices. In: Garrett HE (ed.) North American Agroforestry: an integrated science and practice, 2nd edn. American Society of Agronomy, Madison. 45–55.

Gomez-Baggethun E *et al.* 2010. The history of ecosystem services in economic theory and practice: From early notions to markets and payment schemes. *Ecological Economics*, 69(6): 1209–1218.

Graves AR *et al.* (2007). Development and application of bio-economic modelling to compare silvoarable, arable and forestry systems in three European countries. *Ecological Economics*, 29: 434-449.

Gray LK *et al.* 2016. Climate change risk management in tree improvement programs: selection and movement of genotypes. *Tree Genet. Genomes* 12: 1–12.

Greenberg R *et al.* 1997a. Bird populations inshade and sun coffee plantations in Central Guatemala. *Conservation Biology*,11(2): 448-459.

Greenberg R, *et al.* 1997b. Bird populations in rustic and planted shade coffee plantations of Eastern Chiapas, Mexico. *Biotropica*, 29(4):501-514.

Griffith DM. 2000. Agroforestry: a refuge for tropical biodiversity after fire. *Conserv. Biol.*14(1): 325-326.

Handa AK *et al.* 2007. Effect of pruning intensity on growth and productivity of MPTs and crop under rainfed conditions. *Rainfed Management and Agroforestry* 28(2A): 85-86.

Handa AK *et al.* 2016. Innovative agroforestry for livelihood security in India. World Agriculture, Research gate, 7-15.

Harper JL. 1977. Population Biology of Plants. Academic Press, Orlando, Florida. p892.

Hohne N *et al.* 2012. National GHG emissions reduction pledges and 2 C: comparison of studies. *Climate Policy*, 12: 356-377.

Hutchinson JJ *et al.* 2007. Some perspectives on carbon sequestration in agriculture. *Agric For Meteorol.* 142:288–302.

ICAR. 2010. Degraded and wastelands of India: status and spatial distribution. Indian Council of Agricultural Research, New Delhi, 158p

ICFRE (Indian Council of Forestry Research and Education). 2001. *Forestry Statistics India - 2001*. Dehradun: ICFRE.

IPCC. 2000. Land use, land-use change, and forestry. A special report of the IPCC. Cambridge University Press, Cambridge, pp 375.

IPCC. 2007. Summary for Policymakers. In: Climate Change 2007: Mitigation. Contribution of Working Group III to the Fourth Assessment Report of the Intergovernmental Panel on Climate Change [B. Metz, Davidson, O.R., Bosch, P.R., Dave R. and Meyer L.A. (eds)], Cambridge University Press, Cambridge, United Kingdom and New York, NY, USA.

Johl SS *et al.* 1986. Report of the expert committee on diversification of agriculture in Punjab, submitted to the Govt. of Punjab. 165p.

Jones AT *et al.* 2001. The effect of provenance on the performance of Crataegus monogyna in hedges. *Journal of Applied Ecology*, 38: 952– 962

Jose S. 2009. Agroforestry for ecosystem services and environmental benefits: an overview. *Agroforest Syst*, 76: 1–10.

Karlen DL *et al.* 1997. Soil quality: A concept, definition, and framework for evaluation. Soil Science Society of America Journal, 61: 4-10. https://doi.org/10.2136/sssaj1997.03615995006100010001x.

Katyal JC *et al.* 2016. Global review on state of soil health. *Bulletin of the Indian Society of Soil Science*, 30: 1-33.

Kaushal R *et al.* 2017. Effect of different canopy management practices on rainfallpartitioning in Morus alba. *Ecological Engineering*, 102: 374–380.

Khybri ML *et al.* 1992. Crop yields of rice and wheat grown in rotation as intercrops with tree species in the outer hills of Western Himalayas. *Agroforestry Systems*, 17: 193–204.

Kibblewhite MG *et al.* 2008. Soil health in agricultural systems. *Philosophical Transactions of the Royal Society B Biological Sciences*, 363: 685–701.

Kima DG *et al.* 2016. Carbon sequestration and net emissions of CH4 and N2O under agroforestry: Synthesizing available data and suggestions for future studies. *Agriculture, Ecosystems and Environment* 226 :65–78.

Klaa K *et al.* 2005. Distribution of small mammals in a silvoarable agroforestry in Northern England. *Agroforestry Systems*, 63: 101-110.

Klaus Lorenz and Lal R. 2014. Soil organic carbon sequestration in agroforestry systems. A review. *Agronomy for Sustainable Development*, 34: 443-454.

Koussios S. 2000. Preliminary design of a high speed tumble winder". Delft University of Technology, Department of Aerospace Engineering, Delft.61p.

Kremen C. 2005. Managing ecosystem services: what do we need to know about their ecology? *Ecology Letters*, 8(5):468-479.

Kumar A *et al.* 1990 1991.Tree canopy inter-crop relationships in an agro forestry system.Report 1990 91.Karnal, Haryana: Central Soil Salinity Research Institute.

Kumar K *et al.* 2008. Soil microbial biomass and respiration in afforested calciorthent. *Indian Journal of Agroforestry*, 10(2): 75-83.

Kumar M *et al.* 2018. Influence of *Prosopis cineraria* L. on microclimate conditions under agroforestry system in arid region, India. *Indian J. of Agroforestry*, (20) 1: 35-39.

Kumar BM. 2006. Agroforestry: the new old paradigm for Asian food security. *J. Tropical Agriculture*, 44 (1-2): 1- 14.

Lal R. 1999. Global carbon pools and fluxes and the impact of agricultural intensification and judicious land use. p. 44-52. In Prevention of Land Degradation, Enhancement of Carbon Sequestration and Conservation of Biodiversity through Land Use Change and Sustainable Land Management with a Focus on Latin America and the Caribbean. FAO, Rome, Italy.

Lal R. 2005. Forest soils and carbon sequestration. Forest Ecology and Management. 220 (1-3).

Lande R and Shannon S. 1996. The role of genetic variation in adaptation and population persistence in a changing environment. *Evolution*, 50: 434–437.

Larsen, WE and Pierce FJ. 1994. The dynamics of soil quality as a measure of sustainable management. In: Doran, J.W., Coleman, D.C., Bezdicek, D.F. and steward, B.A. (eds). Defining soil quality for sustainable environment. Soil Science Society of America, special publication no.35,Madison, Wisconsin. Pp.37-51

Lavelle P, Senapati BK and Barros E. 2003. Soil macrofauna. In G. Schroth & F. L. Sinclair (Eds.). Trees, crops and soil fertility: concepts and research methods. Wallingford, UK: CAB International, pp303–323.

Ledig FT and Kitzmiller J. 1992. Genetic strategies for reforestation in the face of global climate change. *For Ecol. Manage.* 50: 153–169.

Lee KH and Jose S. 2003. Soil respiration and microbial biomass in a pecan–cotton alley cropping system in southern USA. *Agrofor Syst*, 58:45–54.

Lee KH, Isenhart TM and Schultz RC. 2003. Sediment and nutrient removal in an established multi-species riparian buffer. *J Soil Water Conserv*, 58:L 1–8.

Long AJ and Nair PKR. 1999. Trees outside forests: agro, community, and urban forestry. *New Forests*, **17**: 145–174.

Lundgren BO and Raintree JB. 1982. Sustained agroforestry. In: Nestel, B. (ed.). Agricultural Research for Development: Potentials and Challenges in Asia, pp. 37-49. ISNAR, The Hague, The Netherlands.

Maikhuri RK, *et al.* 2000. Growth and ecological impacts of traditional agroforestry tree species in Central Himalaya, India. *Agroforestry Systems*, 48: 257-271.

Makundi WR and Sathaye JA. 2004. GHG mitigation potential and cost in tropical forestry-relative role for agroforestry. *Environment, Development and Sustainability*, 6: 235-260.

Manikandan K and Prabhu S. 2013. Agroforestry, Social forestry, urban forestry and joint forest management. **In**: Indian forestry. 2nd edition. Jain brothers, New Delhi.

Manning AD *et al.* 2009. Scattered trees: A complementary strategy for facilitating adaptive responses to climate change in modified landscapes? *Journal of Applied Ecology*, 46: 915-919.

Mátyás C. 1996. Climatic adaptation of trees: rediscovering provenance tests. Euphytica 92: 45. https://doi.org/10.1007/BF00022827

Mazanec RA and Mason ML. 1993. Genetic variation in *Eucalyptus diversicolor* F. Muell. in western Australia and potential gains from selection. *Forest Ecology and Management*, 62: 285–301

McAdam J *et al.* 2007. Developing silvopastoral systems and their effects on diversity of fauna. *Agroforestry Systems*, 70: 81-89.

Mead DJ *et al.* 1999. Farm forestry and shelter. *New Zealand Pasture and Crop Science*, eds. J. White and J. Hodgson, 269–291. Auckland: Oxford University Press.

Mead DJ. 2009. *Agroforestry*. In: Forests and forest plants. Owens, J N and Lung HG (Ed.) Golden Bay, New Zealand. I: 355p.

Meena MB. 2008. Effect of different pruning intensities and doses of fertilizers on the growth and productivity of wheat under Agrisilviculture (Shisham+Wheat) practice. M.Sc.Thesis submitted to the Department of Agronomy, College of Agriculture, JNKVV, Jabalpur.

Mehta S *et al.* 2012. Effect of pruning on Guava CV. Sardar Under Ultra High Density Orcharding System. ICAR Research complex for Eastern Region, Research Centre, Ranchi 25(2): 192-195.

Michel GA *et al.* 2007. Silvopasture for reducing phosphorus loss from subtropical sandy soils. *Plant Soil*, 297: 267–276.

Millennium Ecosystem Assessment. 2003. Ecosystems and Human Well-being: a Framwork for Assessment. Washington D.C. Island Press.

Montagnini F, Francesconi W and Rossi E. 2011. Agroforestry as a tool for landscape restoration. In Agriculture Issues and Policies (USA), Nova Science, New York, p201.

Montgomery DC. 2004. *Design and analysis of experiments*. John Wiley, New York.

Mungai NW *et al.* 2005. Spatial variation of soil enzyme activities and microbial functional diversity in temperate alley cropping systems. *Biol. Fertil. Soils*. 42:129-136.

Nair PKR.1993. An introduction to agroforestry. Kluwer Academic Publishers, Dordrecht, The Netherlands.499p.

Nair PK. 1993. An introduction to agroforestry. Kluwae Acadmic Publishers.

Nair PKR *et al.* 2008. Agroforestry. In: Jorgenson SE, Faith BD (eds.) Encyclopedia of ecology. Elsevier, Oxford. pp 101–110.

Nair PKR *et al.* 2009. Agroforestry as a strategy for carbon sequestration. *J Plant Nutr Soil Sci*. 172:10–23.

Nair PKR and Nair VD. 2003. Carbon storage in North American agroforestry systems. In: Kimble JM, Heath LS, Birdsey RA, Lal R (eds.) The potential of U.S. Forest soils to sequester carbon and mitigate the greenhouse effect. CRC Press, Boca Raton. pp 333–346.

Nair PKR. 2011. Agroforestry Systems and Environmental Quality: Introduction. *J. Environ. Qual.*, 40: 784–790.

Nair PKR. 1993. An Introduction to Agroforestry. Kluwer Academic Publishers, Dordrecht, The Netherlands, pp. 499.

Nair PKR and Latt CR. 1997. Directions in tropical agroforestry research. Agrofor Syst, 38,1–249.

Nair PKR *et al.* 2009. Agroforestry as a strategy for carbon sequestration. *J Plant Nutr Soil Sci*, 172: 10–23.

Nair PKR *et al.* 2010. Carbon sequestration in agroforestry systems. Adv. Agron., 108, 237–307.

Nair VD and Graetz DA. 2004. Agroforestry as an approach to minimizing nutrient loss from heavily fertilized soils: The Florida experience. *Agrofor. Syst.* 61: 269–279.

Nandi A *et al.* 2014. Estimates of the economic contributions of the bidi manufacturing industry in India. *Tobacco contro*:1–8

Nelson ND. 1984. Woody plants are not inherently low in photosynthetic capacity. *Photosynthesis,* 18: 600-605

Noble IR and Dirzo R. 1997. Forests as human-dominated ecosystems. Science, 277, 522–525.

Noda Y *et al.* 2007. Effect of two pruning frequencies and heights on biomass of mulberry. *Zooteenia Tropical* 25(4): 261-268.

NRCAF. 2013. NRCAF Perspective Plan Vision 2050. National Research Centre for Agro-forestry, Jhansi, p. 15-19.

NRM .2006. Watersheds in India: An overview, 1-11.

Oliveira SPD *et al.* 2015. Organic carbon and nitrogen stocks in soils of northeastern Brazil converted to irrigated agriculture. *Land Degrad. Develop.* 26: 9–21.

Ong CK *et al.* 1996. Principles of resource capture and utilization of light and water. In: Ong C.K. and Huxley P. (eds.), Tree-crop interaction. A physiological approach. CAB International, Wallingford, UK.

Palma JHN *et al.* (2007. Methodological approach for the assessment of environmental effects of agroforestry at the landscape scale. *Ecological Engineering*, 29: 450 462.

Palsaniya DR *et al.* 2012. Effect of pruning on growth and forage production in *Albizia procera* based silvi pastoral system in semi–arid, Bundelkhand, India. *Rangeland Management and Agroforestry* 33(1): 96-97.

Pandey DN. 2002. Carbon sequestration in agroforestry systems. *Climate Policy* 2:367–377.

Pandey DN and Mohan D. 1993. Nest site selection by cavity nesting birds on Melia azedarach L. and management of multiple use forests. *J. Bombay nat. Hist Soc.*, 90: 58–61.

Pandey DN. 2007. Multifunctional agroforestry systems in India. *Current Sciences*, 92(4): 455-463.

Panwar VP and Gupta MK. 2015. Litter fall, nutrient return and soil fertility under Celtis australis: an indigenous agroforestry tree species in Himachal Pradesh, India. *International Journal of Environmental Sciences*, 6(3): 318-329.

Parthiban KT *et al.* 2014. Industrial Agroforestry: perspectives and prospective. Scientific Publisher; Jodhpur; New Delhi: India. 384p.

Paustian K *et al.* 2000. Management options for reducing $CO_2$ emissions from agricultural soils. *Biogeochemistry*, 48: 147-163.

Peng RK *et al.* 1993. Diversity of airborne arthropods in a silvoarable agroforestry system. *Journal of Applied Ecology*, 30: 551-562.

Perfecto I *et al.* 1996. Shade coffee: a disappearing refuge for biodiversity. *BioScience*, 46(8): 598-608.

Petit LJ *et al.* 1999. Bird communities of natural and modified habitats in Panama. *Ecography*, 22: 292-304.

Phillips DS *et al.* 1994. Responses of crop pests and their natural enemies to an agroforestry environment. *Agroforestry Forum*, 5(2): 14-20.

Pinney A. 1991. Farmers augmented designs for participating agroforestry research. *Agroforestry Systems*, 15(3): 259-274

Poulsen G. 1983. Using Farm Trees for Fuel wood (in Africa). *Unasylva (FAO)*, 35 (141): 26-29.

Pradhan UB *et al.*. 2003. Effect of frequency of pruning and tree spacing *Luecaena leucocephala* on the growth and productivity of turmeric grown under alley cropping system with Lueacaena. *Journal of Interacademicia*, 7(1): 85-86.

Pulp and paper industry. 2014. A report on opportunities for green chemistry initiatives: Office of the principal scientific adviser to the GOI. Vigyan bhawan. New Delhi..pp 4-7.

Quandt A *et al.* 2017. The role of agroforestry in building livelihood resilience to floods and drought in semiarid Kenya. *Ecology and Society*, 22(3):10-15

Rawat, GS. 2001. Production and marketing of oleo-resin in India. Market Monitoring of Tree Products. Dehradun: Indian Council of Forestry Research and Education.

Rigueiro-Rodríguez A *et al.* 2011. Afforestation of agricultural land with Pinus radiata D. Don and Betula alba L. in NW Spain: effects on soil pH, under storey production and floristic diversity eleven years after establishment. *Land Degrad Dev.* DOI:10.1002/ldr.1072

Rijswijk KV and Brouwer WD. 2001. Application of natural fibre composites in the Development of Rural Societies. Faculty of Aer ospace Engineering; Delft University of Technology, p17.

Rillig MC. 2004. Arbuscular mycorrhizae and terrestrial ecosystem processes. *Ecology Letters*, 7: 740–54.

Rizvi RH *et al.* 2014. Mapping agroforestry in India through remote sensing and prelimary estimates. *Indian farming*, 63(11): 62-64

Rois-Díaz M *et al.* 2006. Biodiversity indicators on silvopastoralism across Europe. Available at http://www.efi.int/files/attachments/publications/tr_21.pdf.

Rousseau GX *et al.* 2012. Indicating soil quality in cacao-based agroforestry systems and old-growth forests: the potential of soil macrofauna assemblage. *Ecological Indicators*, 23: 535-543.

Saigal S and Bose S. 2003. Small-scale forestry enterprises in India: overview and key issues.Winrock International India, New Delhi; International Institute for Environment and Development:London. 63p.

Sarkar PK *et al.* 2017. Improvement in agroforestry system. *Indian farming*, 67(7): 19-20.

Sarvade S *et al.* 2014. Wheat (Triticum aestivum L.) Yield and soil properties as influenced by different agri-silviculture systems of Terai Region, Northern India. *International Journal of Bio-resource and Stress Management*, 5(3): 350-355.

Schroeder P. 1994. Carbon storage benefits of agroforestry systems. *Agroforest Syst*. 27:89–97.

Schroeder P. 1994. Carbon storage benefits of agroforestry systems. *Agroforest Syst*, 27: 89-97.

Schroth, G. & Sinclair, F.L. (2003) (eds) Trees, Crops and Soil Fertility: Concepts and Research Methods. CABI Publishing, Wallingford.

Seiter S *et al.* 1995. Increase in soil microbial biomass and transfer of nitrogen from alder to sweet corn in an alley cropping system. In: J.H. Ehrenreich., D.L. Ehrenreich, & H.W. Lee (Eds) Growing a sustainable future. University of Idaho, Boise, pp 56–158.

Semwal RL *et al.* 2003. Leaf litter decomposition and nutrient release patterns of six multipurpose tree species of central Himalaya, India. *Biomass Bioenergy*, 24: 3-11.

Sen PP. 1984. Potential of planting *Leucaena leucocephala* on the marginal land at Titta village Ranchi, Annual Extension Report, Faculty of Forestry, B.A.U. Ranchi, pp.1-8.

Sharma JP *et al.* 2014. Enhancing Farm Profitability Through Food Processing and Value Addition. In: Shukla, J.P. Technologies for Sustainable Rural Development: Having Potential of Socioeconomic Upliftment (TSRD-2014). Published by allied publisher Pvt.Ltd., India. p. 88.

Sharrock RA *et al.* 2004. A global assessment using PCR techniques of mycorrhizal fungal populations colonising Tithoniadiversifolia.*Mycorrhiza*, 14: 103–9.

Sharrow SH and Ismail S. 2004. Carbon and nitrogen storage in agroforests, tree plantations, and pastures in western oregon, USA. *Agroforestry Systems*, 60: 123-130.

Sharrow SH and Ismail S. 2004. Carbon and nitrogen storage in agroforests, tree plantations, and pastures in western Oregon, USA. *Agroforest Syst*, 60: 123–130.

Shastri CM *et al.* 2002. Tree species diversity in a village ecosystem in Uttara Kannada district in Western Ghats, Karnataka. *Curr. Sci.*, 82: 1080-1084.

Shaw RP. 2008. Impact of population growth on Environment: The debate heats up. *Elsevier*, 12(1-20), 11-36.

Sidaa TS *et al.* 2018. Climate-smart agroforestry: *Faidherbia albida* trees buffer wheat against climatic extremes in the Central Rift Valley of Ethiopia. *Agricultural and Forest Meteorology* 248: 339–347

Sileshi G *et al.* 2007. Contributions of agroforestry to ecosystem services in the Miombo eco-region of eastern and southern Africa. *Afr J Environ Sci Technol.* 1: 68–80.

Singh C *et al.* 2008. Evaluation of management practices for a *Grewia optiva*-hybrid napier-based silvi-pastoral system for rehabilitating old riverbed lands in the North-Western Himalaya. International Journal of Ecology and Environmental Sciences, 34(4): 319-327. Fischer KS and Palmer AFE. 1984. Tropical maize. The physiology of Tropical field Crops. Pp. 123-248.

Singh NB *et al.* 2001. All India coordinated research project on poplar improvement: need, implementation and achievement. *Advances in Horticulture and Forestry*, 8: 205-237.

Singh S. 2013. Praj Industries Revs up Cellulosic Ethanol Plan. India Frobes, India.

Singh S. 1994. Physiological response of different crop species to light stress. *Indian Journal of Plant Physiology* 37: 147-151.

Singh AN *et al.* 2002. Plantation as a tool for mine spoil restoration. Curr Sci, 82:1436–1441.

Singh YP *et al.* 2014. Bio-amelioration of alkali soils through agroforestry systems in central Indo-Gangetic plains of India. *Journal of Forestry Research*, 25(4): 887 896.

Sisay M and Ketema H. 2015. Variation in abundance and diversity of soil invertebrate macro-fauna and some soil quality indicators under agroforestry based conservation tillage and maize based conventional tillage in Southern Ethiopia. *International Journal of Multidisciplinary Research and Development* 2(8): 100-107.

Six J *et al.* 2000. Soil macroaggregate turnover and microaggregate formation: a mechanism for C sequestration under no-tillage agriculture. *Soil Biol Biochem.* 32: 2099–2103.

Smith J *et al.* 2012. Reconciling productivity with protection of the environment: Is temperate agroforestry the answer? *Renewable Agriculture and Food Systems*: 1–13. doi:10.1017/S1742170511000585.

Solomon S *et al.* 2007. The Physical Science Basis. Contribution of Working Group I to the Fourth Assessment Report of the Intergovernmental Panel on Climate Change. Cambridge, UK and New York, NY, USA: Cambridge University Press; 2007, (IPCC Fourth Assessment Report (AR4)). Spain. J Environ Qual. 40:1–8.

Sridhar KR and Bagyaraj DJ. 2017. Microbial Biodiversity in Agroforestry Systems. In: Dagar J., Tewari V. (eds) Agroforestry. Springer, Singapore

Srivastava A *et al.* 2017. Value Addition in Minor Fruits of Eastern India: An Opportunity to Generate Rural Employment. In: Dutta, A. K. and Mondal, B. (Eds.), Fruits for Livelihood: Production Technology and Management Practices Published by Agrobios (India), Jodhpur, India. pp. 395-417.

State of the World's Forests. (2009). Global demand for wood products: 33p.

Stehman SV and Meredith MP. 1995. Practical analysis of factorial experiments in forestry. *Canadian Journal of Forest Research* 25(3): 446-461

Swain KC. Biofuel Production in India: Potential, Prospectus and Technology. Department of ASEPAN, Institute of Agriculture, Visva-Bharthi University Birbhum, West Bengal, India Volume 4 (1):2-4.

Tandon JC. 1991. The safety match industry in India. In: Campbell, J.Y. (Ed.) Case Studies in Forest-Based Small-Scale Enterprises in India. Bangkok: Food and Agriculture Organisation of the United Nations.

Tapia-Coral SC *et al.* 1999. Litter-layer macrofauna in agroforestry systems on abandoned pastures in central Amazonia. *Acta Amazonica*, 29 (3): 447–495.

TAU Agritech Portal. 2016. Watershed Development Programme, Tamil Nadu Watershed Development Agency (TAWDEVA).

Terry. 2000. USDA National Agroforestry Center. Agroforestry extension manual for Kenya. Ebook published by ICRAF,Kenya.

Tewari DN. 1995. Marketing and Trade of Forest Produce. Dehradun: International Book Distributors.

Thapa GB. 2005. Integrated Watershed Management: Basic Concepts and Issues Human Settlement Development, 2, 1-7.

Thrupp LA. 1997. Linking biodiversity and agriculture: Challenges and opportunities for sustainable food security. World Resources Institute, USA.

Thurston HD *et al.* 1999. Traditional management of agrobiodiversity. In D. Wood & J.M. Lenne (Eds) Agroforestry: characterization, utilization and management. CABI, Wallingford, pp 211–243.

Tiwari P *et al.* 2017. Agroforestry for sustainable rural livelihood: A review. *International Journal of Pure & Apllied Bioscience.* 5 (1): 299-309.

Tomar OS and Gupta RK. 1984-94. Performance of tree species in saline soils. Report, CSSRI, Karnal, India.

Tuli DK. 201. Opportunities in 2nd generation biofuel-current status. Petrofed, Indian oil corporation limited, New Delhi, India.

Tyndall J and Colletti J. 2007. Mitigating swine odor with strategically designed shelterbelt systems: a review. *Agrofor Syst.* 69: 45–65.

Udawatta RP *et al.* 2008a. Agroforestry and grass buffer effects on pore characteristics measured by high-resolution X-ray computed tomography. *Soil Sci Soc Am J.* 72: 295–304.

Udawatta RP *et al.* 2008b. Variations in soil aggregate stability and enzyme activities in a temperate agroforestry practice. *Applied Soil Ecology*, 39:153-160.

Udawatta RP *et al.* 2002. Agroforestry practices, runoff, and nutrient loss: a paired watershed comparison. *J Environ Qual*, 31: 1214–1225.

UNEP. 2004. Global Environment Outlook 3: Past, present and future perspectives. United Nations Development Programme, Nairobi, Kenya.

UNFCCC. 2007. Report of the conference of parties on its thirteenth session, Bali, Indonesia.United nations framework convention on climate change, Geneva UN, 1998).

United Nations (UN). 1998. Methodological Issues: Issues related to land-use change and forestry, Note by the Secretariat. FCCC/SBSTA/1998/INF.1, 18 May.

USDA-NRCS. 2018. Healthy soil for life. https://www.nrcs.usda.gov/wps/portal/nrcs/main/soils/ health/. Accessed September 29, 2018.

Uthappa AR *et al.* 2015. Soil physico-bio-chemical properties as influenced by varying tree densities in poplar (*Populus deltoids* Bartr. ex Marsh.) based agroforestry system. *Indian Journal of agroforestry*, 17 (1): 81-90.

Van Noordwijk *et al.* 1996. Root distribution of trees and crops: competition and/or complementarity. In C.K. Ong., Huxley, P. (Eds) Tree–Crop interactions: a physiological approach (pp 319–364), CAB International, Wallingford, UK.

Van A *et al.* 2003. Poverty and climate change: Reducing the vulnerability of the poor through Adaptation. Paris: Organisation for Economic Co-operation and Development.

Wallace KJ. 2007. Classification of ecosystem services: problems and solutions. *Biological Conservation*, 139: 235-246.

Wardle DA *et al.* 2004. Ecological linkages between aboveground and belowground biota. *Science*, 304: 1629–33.

Waston TT *et al.* 2000. Land use, Land-use Change and Forestry, IPCC Special Report. Cambidge University Press.

Watson RT *et al.* 2000. Land Use, Land-Use Change and Forestry. IPCC Special Report, Cambridge Univ. Press, Cambridge, pp 388.

Werf WVD *et al.* 2007. Yield-safe: A parameter-sparse process-based dynamic model for predicting resource capture, growth and production in agroforestry systems. *Ecological Engineering*, 29: 419-433.

Wiersum F and Dirjosoemarto S. 1987. Past and Current Research with Gliricidia in Asia, *In*: Gliricidia sepium management and improvement. N.F.T.A., Hawai.

Williams-Guille *et al.* 2008. Bats limit insects in a tropical agroforestry system. *Science*, 320: 58-72.

Winterbottom R and Hazleword PT. 1987. Agroforestry and sustainable development making the connection. *Ambio*, 6(2-3): 100-110.

Wooding WN. 1973. The split plot design. *Journal of Quality Technology,* 5: 16-33

Wright C. 1994. The distribution and abundance of small mammals in a silvoarable agroforestry system. *Agroforestry Forum,* 5(2): 26-28.

Youden WJ. 1972. Randomization and experimentation. *Technometrics* 14: 13-22

Young A. 1997. *Agroforestry for Soil Management.* Wallingford: CAB Internationa. 320 pp

Zhang Q *et al.* 2018. Impact of tree species on barley rhizosphere-associated fungi in an agroforestry ecosystem as revealed by 18S rDNA PCR-DGGE. Agrofor Syst. 92(2): 541-554.

Zobel BJ *et al.* 1958. Seed orchards, Their concept and management. *J For* 56: 815-825

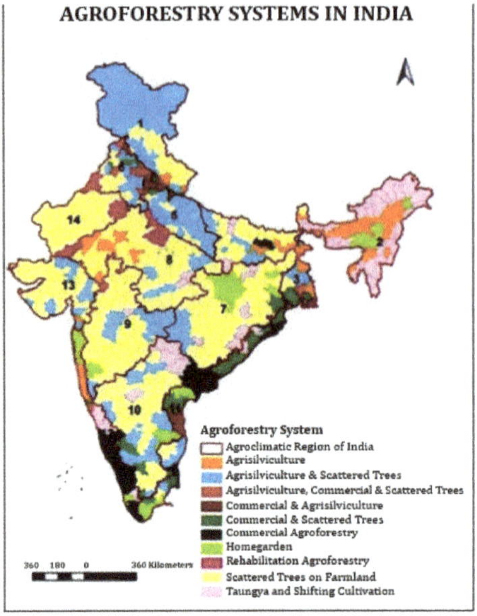


**Figure 4. Thematic map showing various agroforestry systems in India (Source: Rizvi et al., 2014) p. 11**

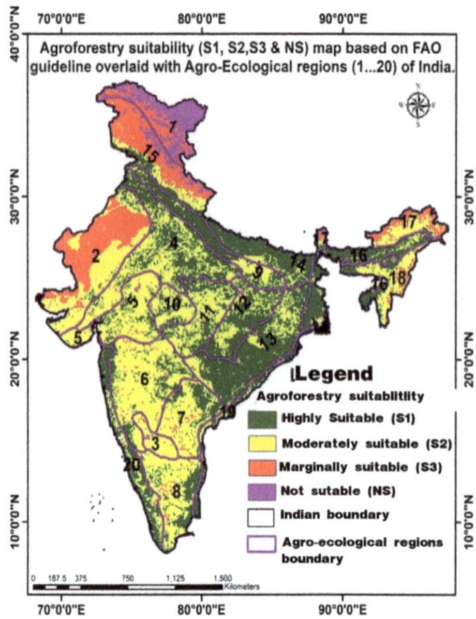


**Figure 5. Agroforestry suitability map of India (Source, Ahmed et al., 2018) p. 12**

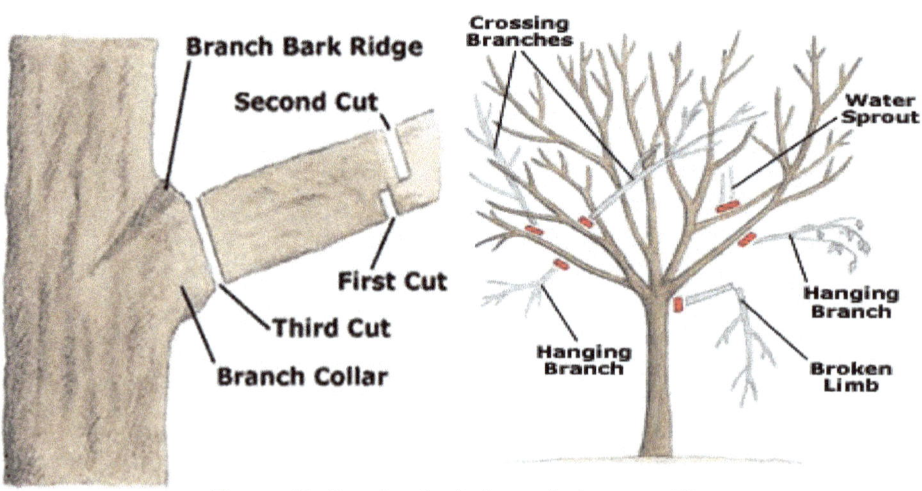


**Figure 13. Pruning techniques in trees. p. 52**

**Figure 19: Wheat with *Populus deltoides* Turmeric (*Curcuma domestica*) with *Populus deltoides* p. 71**

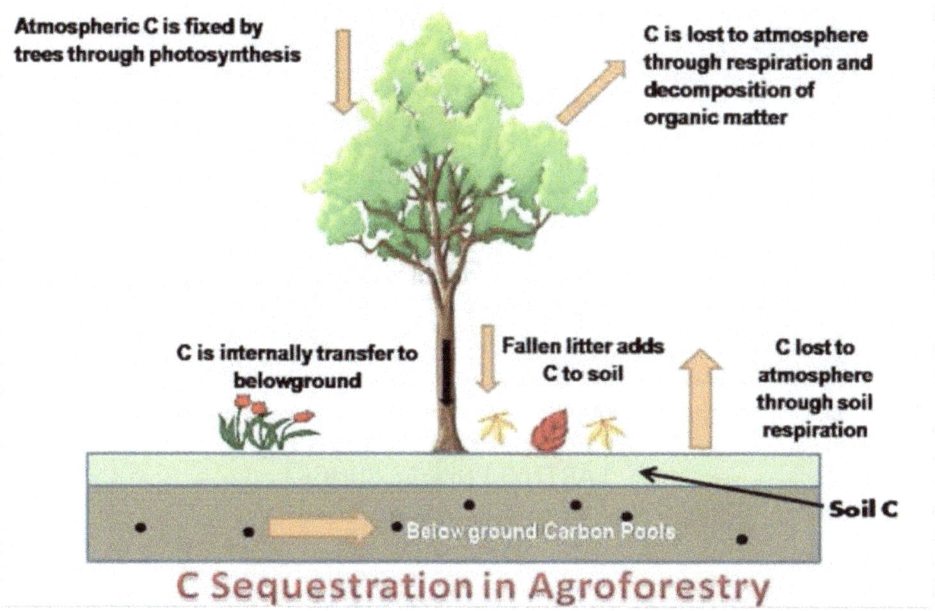


**Figure 21. Carbon sequestration process in agroforestry (Source Newaj et al., 2014) p. 78**

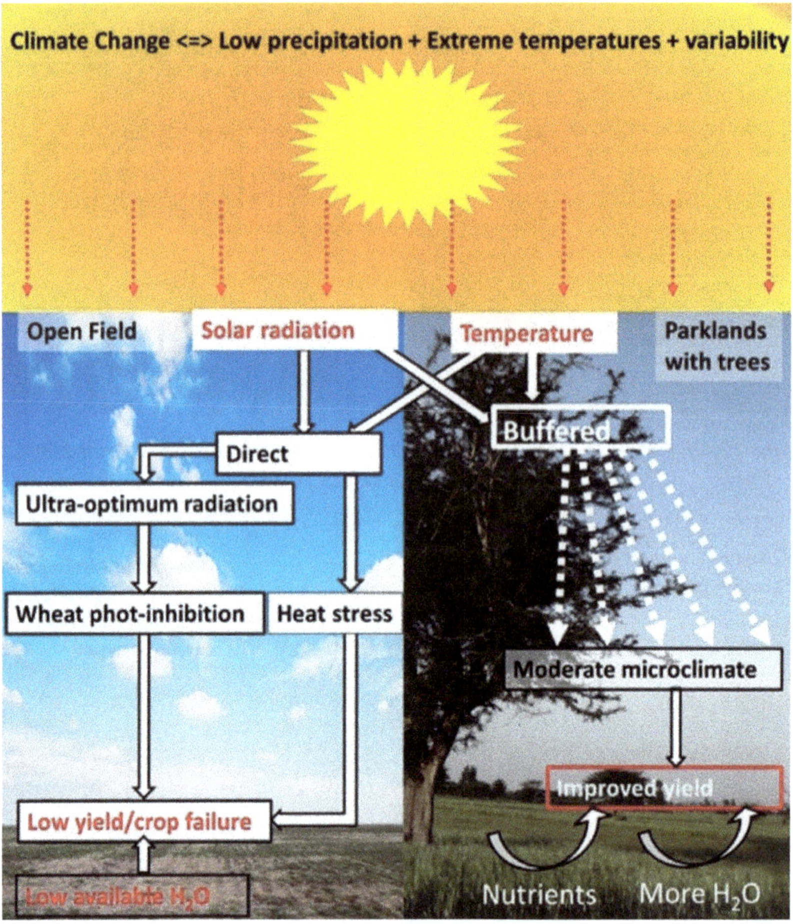


**Figure 22. Climate change mitigation potential of wheat based agroforestry systems (Source: Sidaa et al. 2018) p. 79**

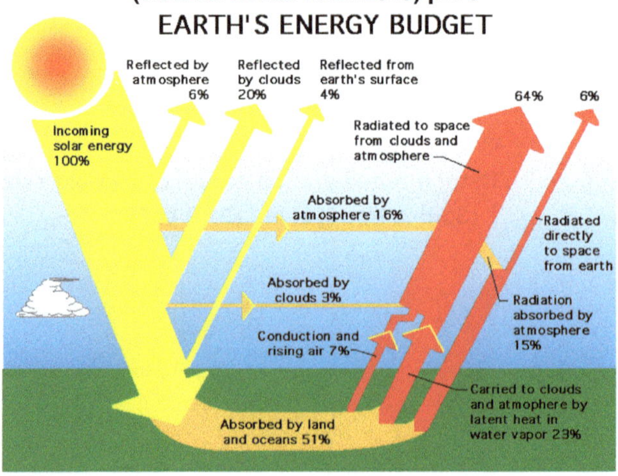


**Figure 32. Effects of clouds on the Earth's Energy Budget (image is from a NASA site) p. 111**

**Figure 39. Mechanical planter p.159**

**Figure 40. Pitting machine p.160**

**Figure 41. Forwarder p. 160**

**Figure 42. Harvester p. 161**

**Figure 42. Harvester p. 161**

**Figure 41. Forwarder p. 162**

www.ingramcontent.com/pod-product-compliance
Ingram Content Group UK Ltd.
Pitfield, Milton Keynes, MK11 3LW, UK
UKHW021322190726
13845UKWH00003B/22

9 789390 371440